Library of
Davidson College

THE ORIGIN OF THE STRATEGIC CRUISE MISSILE

THE ORIGIN OF THE STRATEGIC CRUISE MISSILE

Ronald Huisken

PRAEGER

PRAEGER SPECIAL STUDIES • PRAEGER SCIENTIFIC

Library of Congress Cataloging in Publication Data

Huisken, Ronald.
   The origin of the strategic cruise missile.

   Includes bibliographical references and index.
   1. Cruise missiles—History. 2. Atomic weapons
—History. I. Title.
UG1312.C7H84   355.8'2519   81-4921
ISBN 0-03-059378-6   AACR2

Published in 1981 by Praeger Publishers
CBS Educational and Professional Publishing
A Division of CBS, Inc.
521 Fifth Avenue, New York, New York 10175 U.S.A.

© 1981 by Praeger Publishers

All rights reserved

123456789 145 987654321

Printed in the United States of America

TO
MIELING
JAN AND MARC

ACKNOWLEDGMENTS

The two people to whom I owe the most are my friends and former supervisors, Dr. Robert O'Neill and Dr. Desmond Ball of the Strategic and Defence Studies Centre, Australian National University.

Although the research was not based to any great extent on interviews, I would like to acknowledge the following persons with whom I had fruitful discussions: Dr. Stu Rubens of the U.S. Department of Defense, Col. Ryan (USAF) of the U.S. Arms Control and Disarmament Agency, Dr. F. S. Nyland of the Rand Corporation, and Dr. Freeman Tatum, also of the Rand Corporation. I would like to thank the United States Air Force for permission to use the Albert F. Simpson Historical Research Center at Maxwell Air Force Base in Montgomery, Alabama.

I also have to thank Professor Warner Schilling, Professor Michael Nacht, Professor Jack Ruina, and Dr. Frank Barnaby who, as examiners of the thesis, made a number of helpful comments and criticisms. Finally, I am grateful to Ms. Carmencita Reyes for typing the manuscript. None of the above, however, can be held in any way responsible for what follows. This is also true of the United Nations, with whom I happened to be employed at the time this book went to press.

CONTENTS

ACKNOWLEDGMENTS   vii

ABBREVIATIONS AND ACRONYMS   xi

INTRODUCTION   xiii

PART I: PROGRAMMATIC HISTORY

Chapter

1. TECHNICAL CHARACTERISTICS OF MODERN CRUISE MISSILES   3

    Notes   13

2. U.S. LONG-RANGE CRUISE MISSILES: A HISTORICAL NOTE   15

    Introduction   15
    Early Strategic Cruise Missiles   16
        Snark and Navaho   16
        Hound Dog   18
        Regulus   20
    Early Tactical Cruise Missiles   21
    Notes   24

3. THE SUBMARINE-LAUNCHED CRUISE MISSILE: A WEAPON IN SEARCH OF A MISSION   28

    Introduction   28
    The Strategic Cruise Missile Proposal   31
    The Official Rationale   32
    Initial Reactions in Congress   36
    Subsequent Developments   39
    Notes   53

4. THE AIR-LAUNCHED CRUISE MISSILE: PENETRATING VERSUS STANDOFF STRATEGIC BOMBERS   60

    Introduction   60
    The Armed Decoy Cruise Missile   64

	Enter the ALCM	71
	Notes	87

PART II: EVALUATING THE DEVELOPMENT OF CRUISE MISSILES

	Introductory Note	96
5	THE MILITARY REQUIREMENT FOR STRATEGIC CRUISE MISSILES	97
	Soviet Strategic Cruise Missiles	97
	Cruise Missiles and the Wider Strategic Balance	104
	Weighing the Balance	109
	Debating the Balance	117
	Conclusion	127
	Notes	129
6	STRATEGIC DOCTRINE: COUNTERFORCE, LIMITED OPTIONS, AND THE CRUISE MISSILE	135
	Introduction	135
	Soviet Strategic Doctrine	142
	The Schlesinger Doctrine	147
	After Schlesinger	153
	Assessment	157
	Notes	160
7	TECHNOLOGICAL MOMENTUM	166
	Notes	172
8	SALT	174
	Notes	183
9	CONCLUSIONS	186
	Notes	192
EPILOGUE		193
APPENDIX: U.S. Congressional Hearings Cited in Text		195
INDEX		201
ABOUT THE AUTHOR		203

ABBREVIATIONS AND ACRONYMS

ABM	Antiballistic Missile
ALCM	Air-Launched Cruise Missile
ASW	Antisubmarine Warfare
AWACS	Airborne Warning and Control System
CEP	Circular Error Probable; radius of a circle centered on the target within which 50 percent of a large number of warheads would fall
CNO	Chief of Naval Operations
DDR&E	Director of Defense Research and Engineering
DSARC	Defense Select Acquisition Review Council
FBS	Forward Based Systems
FY	Fiscal Year
GLCM	Ground-Launched Cruise Missile
ICBM	Intercontinental Ballistic Missile
IOC	Initial Operational Capability
IRBM	Intermediate-Range Ballistic Missile
KT	Kiloton; explosive equivalent of 1,000 tons of TNT
MAP	Multiple Aim Point; a basing system in which a given number of ICBMs would be moved at random between a much larger number of launch silos
MIRV	Multiple Independently Targetable Reentry Vehicles
MRBM	Medium-Range Ballistic Missile
MT	Megaton; explosive equivalent of one million tons of TNT
NATO	North Atlantic Treaty Organization
n.m.	nautical mile
OSD	Office of the Secretary of Defense
PBV	Postboost Vehicle; a ballistic missile payload capable of maneuvering after the boost phase (together with a warhead-dispensing mechanism, this constitutes a MIRV system)
psi	pounds per square inch; overpressure, expressed in psi, is the measure commonly used to determine the capability

	of an object to withstand the pressure exerted by a nuclear blast
QRA	Quick Reaction Alert; tactical nuclear weapons systems maintained on a high state of alert
RCS	Radar Cross Section
R&D	Research and Development
RDT&E	Research, Development, Test, and Evaluation
RV	Reentry Vehicle; that portion of a ballistic missile designed to carry a nuclear warhead and reenter the earth's atmosphere in the final stage of a missile's trajectory
SALT	Strategic Arms Limitation Talks
SAM	Surface-to-Air Missile
SCAD	Subsonic Cruise Armed Decoy
SCM	Strategic Cruise Missile
SIOP	Single Integrated Operational Plan or Strategic Integrated Operations Plan
SLBM	Submarine-Launched Ballistic Missile
SLCM	Sea-Launched or Submarine-Launched Cruise Missile
SMAC	Scene-Matching Area Correlator
SRAM	Short-Range Attack Missile
SSBN	Nuclear-powered submarines armed with long-range ballistic missiles
SSN	Nuclear-powered attack submarine
TALCM	Tomahawk Air-Launched Cruise Missile
TERCOM	Terrain Contour Matching
TNW	Tactical Nuclear Weapon
ULMS	Undersea Long-range Missile System
WTO	Warsaw Treaty Organization

INTRODUCTION

For over two decades the weapons acquisition process has attracted a great deal of study and debate. This interest is well placed. The development and production of major weapons is an extremely serious business, and commonsense demands the most vigorous scrutiny to insure a maximum of rationality and a minimum of weapons. What has emerged most clearly from all this work is that rational behavior in this field is particularly hard to define and even harder to enforce. The weapons acquisition process is a most complex amalgam of political, military, technological, economic, and bureaucratic considerations. Experience has shown that it is not necessary for all these factors to pull together in order for a new weapons system to survive through to operational deployment. Indeed, it is quite rare for this to happen. Past studies suggest that some subset of these factors can usually account adequately for the emergence of a new weapon, with the result that any conclusion regarding rationality becomes a judgment on the part of the analyst.

Experience has also shown that the longer a weapon system survives through concept studies, development, and testing, the more comprehensive the support for it becomes and the more difficult it is to stop. Vested bureaucratic and economic interests grow over time. Similarly, the probability that the adversary will initiate a similar effort increases with time; once this happens, going ahead with the weapon makes sound military and political sense particularly, of course, for a major nuclear system.

In investigating the history of a major weapons system it is useful to distinguish between the original decision to launch a development program and the subsequent vicissitudes of the program. This is especially true if the system in question is conceptually new rather than just a more technologically advanced follow-on to an existing system. In the strategic nuclear arena, examples of conceptually new systems include the first intercontinental ballistic missile (ICBM), the submarine-launched ballistic missile (SLBM), multiple independently targetable reentry vehicles (MIRV), and, in the author's opinion, the modern strategic cruise missile.

Two considerations suggested that the story of the strategic cruise missile might be somewhat atypical and therefore worth looking into. The first was that the proposal to develop a cruise missile for strategic use meant U.S. thinking on this type of weapon had come full circle. The strategic cruise missile was thoroughly explored in the 1950s and decisively abandoned in favor of ballistic missiles. Moreover, subsequent technological developments in range, reliability, and accuracy had merely reinforced the superiority of the ballistic missile. In other words, it was far from obvious what advantages a small, subsonic, and, relatively speaking, short-range cruise missile could offer in the strategic role. The second consideration was that the strategic cruise missile made a surprisingly abrupt debut. The development of a new strategic weapons system in the United States is normally preceded by an extended period of open debate on the requirement for the weapon, its performance characteristics, force size, and so on. This was the case, for example, with MIRV, the B-1 strategic bomber, and the MX ICBM. In contrast, there was no open debate at all on the strategic cruise missile prior to the urgent request that one be developed. In fact, even within the defense establishment the gestation period of the proposal to develop this weapon was remarkably short.

The history of the modern strategic cruise missile actually involves two weapons, one submarine launched (SLCM) and the other bomber-launched (ALCM). The histories of these two weapons are closely interrelated, which is a complicating factor so far as the arrangement of the following discussion is concerned. For convenience and clarity, a separate chapter will be devoted to the development history of each weapon. As regards the analysis of why the strategic cruise missile staged its remarkable comeback, the distinction made above will be followed, namely, the reasoning behind the original proposal to develop such a weapon on the one hand, and its subsequent history through to the deployment decision on the other. As it happens, the original proposal concerned the SLCM exclusively while the strategic cruise missile that will be deployed is the ALCM. Accordingly, the analysis of the history of this weapon will be centered on how and why it was transformed from a submarine-launched weapon to one launched from bombers.

The history of the modern cruise missile is clearly just beginning. Current plans call for the production of 3,418 AGM-86B ALCMs over the period 1980-89 for deployment on approximately 150 B-52G strategic bombers; ground- and submarine-launched cruise missiles are under development as theater nuclear weapon systems; and submarine-, ship-, and air-launched variants are under development for the tactical antishipping role. For the present purpose, however, the key period extends from June 1972, when development of the SLCM was first proposed, to June 1977, when President Carter canceled the B-1 in favor of a large force of ALCMs to be deployed on existing strategic bombers. More recent developments are incorporated only if they are particularly germane to the discussion.

THE ORIGIN OF THE STRATEGIC CRUISE MISSILE

I
PROGRAMMATIC HISTORY

1

TECHNICAL CHARACTERISTICS OF MODERN CRUISE MISSILES

A cruise missile can be defined as a self-guided, dispensable weapon system that operates in sustained aerodynamic flight propelled continuously by an air-breathing engine. This definition distinguishes a cruise missile from all ballistic missiles since the latter are neither continuously powered nor continuously guided, and most leave the atmosphere during some portion of their flight. It also distinguishes a cruise missile from weapons like the Soviet SSN-2 Styx and the French MM-38 Exocet antiship missiles. Both these weapons fly through the atmosphere and are continuously propelled, but the propulsion system is a rocket motor and their wings serve mainly to provide stability and control rather than aerodynamic lift.

The two weapons central to this study are the air-launched cruise missile (ALCM) and the sea-launched cruise missile (SLCM). The SLCM, designated BGM-109 Tomahawk, is being developed in three variants by the Convair Division of the General Dynamics Corporation. The basic version is a submarine-launched, nuclear-armed weapon originally intended for strategic missions but now labeled a theater land-attack weapon. The two derivative weapons are a submarine-launched antiship missile with a conventional warhead, and a land-attack missile with a nuclear warhead launched from cannisters carried on a truck, and usually labeled the ground-launched cruise missile (GLCM).

All three variants have identical external dimensions: 18 feet 3 inches long (or 20 feet 3 inches including the rocket booster), a wing span of 8 feet 7 inches, and a diameter fractionally less than 21 inches to permit launch from standard submarine torpedo tubes. The wings, the air-intake, and the four control surfaces at the rear all fold away so that the missile can be stored and fired from a cannister that fits exactly into a torpedo tube.

A fourth variant of this basic weapon, designed AGM-109, was an entrant in the competition for the ALCM. For this role, the weapon was lengthened to 20 feet but the wing span and diameter remained the same. The winning entrant in the ALCM competition was Boeing's AGM-86B, a descendant of a decoy missile called SCAD (subsonic cruise armed decoy) that was canceled in 1973. The Boeing weapon is 20 feet 9 inches long and has a wing span of 12 feet. In shape, it is more triangular than cylindrical but, like the AGM-109, the wings, the air-intake, and the rear control surfaces fold neatly away for storage on the launcher.

The submarine-launched land-attack version of the SLCM will be carried by nuclear-powered hunter/killer submarines (SSNs). United States Navy studies suggest that an SSN could carry up to six SLCMs without significantly reducing its capacity in the ASW role, which is the primary mission assigned to these units. Under normal circumstances, an SSN would carry two or three land-attack SLCMs with the remaining spaces left for the antiship version. The submarines would operate in areas relevant to their ASW mission until directed to proceed to launch points for the land-attack SLCM. Since this may take anything from an hour to more than a day, the 25 minutes required to prepare the SLCM for launch is of no great consequence.

When the encapsulated missile is in the torpedo tube and ready for launching, it is pushed from its capsule by the submarine's hydraulic ejection system. When the missile is about 30 feet in front of the submarine, a lanyard attached to the capsule is pulled to ignite the rocket booster. When this sequence is complete, the capsule is also ejected from the torpedo tube. The rocket booster, which fires for about 12 seconds, propels the missile through the water at an angle of about 50 degrees and up into the atmosphere to an altitude of 1,300 feet. This allows time for the wings and air-intake to be deployed and for the missile to achieve a velocity adequate for aerodynamic flight (something in excess of Mach 0.44). When the booster burns out it is promptly separated from the missile, a gas cartridge is fired to start the turbofan engine, and the missile flies off in its cruise configuration. Depending on the location and intensity of the defenses the missile will have to penetrate on the way to its target, it can be programmed to climb initially to altitudes that maximize fuel economy (about 20,000 feet) and subsequently descend to very low altitudes before coming within radar detection range. Alternatively, it can descend immediately to low altitude.

The launch sequence for the GLCM is essentially the same except that the booster rocket launches the missile directly from its cannister. During the transition to cruise flight, the weapon reaches a maximum altitude of about 1,000 feet.[1] The booster is jettisoned after the missile has traveled about 4,000 feet, after which it coasts for nearly a mile while the turbofan winds up to full thrust.

The air-launched cruise missile will initially be deployed on B-52G strategic bombers, 173 of which are currently in the active inventory. Eight missiles will be carried internally on a rapid-launch dispenser in the bomb bay. In addition, 12 missiles, in two clusters of six, will be carried on wing pylons.

Launch at high altitude can significantly increase the weapon's range. For example, the range claimed for the original AGM-86A, 690 n.m., assumed a low-level launch and low-level flight for the missile after launch. If the weapon was launched at 45,000 feet, programmed to descend to 26,000 feet, cruise-climb back to 30,000 feet, and finally drop to tree-top level in the terminal phase, a range of 1,000 n.m. was considered possible.[2] Whether or not this is possible depends on whether the enemy can detect and attack the B-52s before they reach optimal cruise missile launch points. If they can, the launching aircraft would have to descend to low altitude before firing their cruise missiles. Since the ALCM is equipped with an oxygen bottle to ease engine start-up at high altitudes, it is clear that the United States expects this luxury to be available for some time, at least for selected launch points.

The ALCM, whether it is carried internally or on the wing pylons, must fall free of the aircraft before the wings and tail surfaces deploy and the engine is started. The wings are fully deployed two seconds after launch and the engine is up to full thrust after five seconds. By the time the ALCM is flying under its own propulsion it will have dropped a total of about 450 feet. The significance of this parameter is that the higher the launching aircraft must fly while releasing its missiles, the more likely becomes its detection by enemy radars.

A number of distinct technological developments in propulsion units, guidance systems, and warhead design made the cruise missile possible, but a common factor was the ability to make things smaller. The viability of the cruise missile as a weapons system depends more on its small size than anything else. Because it is subsonic and cannot do anything but what it is programmed to do, the survivability of the cruise missile depends critically on it remaining undetected until it is too late for the defenses to react. The cruise missile's low penetrating altitude, claimed to be as low as 50 feet over reasonably smooth terrain, makes an important contribution to its survivability, but the really vital parameter is radar cross section (RCS).

Hard official information on RCS is lacking and such indications given in the open literature are sometimes contradictory. The best known statement on this issue is that "we have our missile [RCS] down to seagull size now, and are striving to match the size of a sparrow."[3] One source estimates that the cruise missile has an RCS of just 0.05 square meters, and that this is about 0.05 percent that of a B-1 strategic bomber.[4] Another source suggests that the SLCM's RCS is about 10 percent that of an F-4 Phantom II fighter-bomber.[5] These indications on the relative size of a cruise missile's radar image are not necessarily comparable. A weapon's RCS depends quite heavily on its aspect in relation to the radar, and this has not been specified. The image is smallest in a head-on approach, considerably larger when the target is side-on to the radar, and probably larger still if the radar can look down on the target (provided the confusion caused by ground clutter can be filtered out). In any event, it is clear that the radar image of U.S. cruise missiles is minuscule compared to that of a strategic bomber, a fact that played a key role in the decision to proceed with this weapon rather than the B-1.

Another potential source of danger for any vehicle penetrating enemy airspace at less than hypersonic speeds is the heat energy generated by its engine. Infrared-guided surface-to-air missiles (SAMs) can home in on this heat. Here again the cruise missile protects itself by being small. Its tiny engine presents a very small target for heat-seeking missiles and, because it is a turbofan, the temperature of the exhaust plume is a relatively low 600 degrees Fahrenheit.[6] A recent modification to the AGM-86B was the addition of a boattail (in effect an exhaust) to further reduce the infrared signature.

A small, efficient turbine engine was one of the two major technical advances that made the current long-range cruise missiles possible. All existing variants of long-range cruise missiles in the United States—SLCM, ALCM, and GLCM—have standardized on a turbofan engine developed by the Williams Research Corporation and designated F107-WR-100. The F107 weighs just 130 lbs. and its maximum diameter is 12 inches, but it nonetheless generates some 600 lbs. of thrust.

This engine is also remarkably economical for its size with a specific fuel consumption (SFC) of about 0.7.[7] SFC is a measure of the weight of fuel consumed as a function of both time and the amount of thrust generated. Thus an SFC of 0.7 means that 0.7 lbs. of fuel is consumed per hour per pound of thrust generated. Jet engines can be programmed to give a range of speeds. The ALCM and SLCM apparently must fly at more than Mach 0.44 in order to stay in the air, but between this figure and the sonic barrier speed can be selected to improve survivability (higher speed) or to maximize fuel economy (lower speed).[8] Most sources give Mach 0.7 as the design cruise speed of both the ALCM and SLCM, which is equivalent to 530 mph at sea level.[9]

Fuel economy is also a function of the type of fuel used. Both ALCM and SLCM can use standard aviation fuels like JP-4 and JP-5, but range can be significantly increased if thicker fuels are used. For example, using JP-5 fuel, the range of the air-launched version of the Tomahawk was estimated at 1,600 n.m. In the submarine-launched version, using a fuel called TH-Dimer, the identical missile could achieve a range of 2,100 n.m. The Navy is working on an even thicker fuel called Shelldyne, which could increase the SLCM's range by at least another 20 percent.[10] An important constraint on the use of high density fuels in the ALCM is that these fuels freeze in the low temperatures experienced at high altitudes.

The second key to the efficacy of a long-range cruise missile is the ability to periodically check the accuracy of the weapon's flight toward the target and, when necessary, to correct any tendencies to wander that may have developed. Being subsonic, a cruise missile takes a relatively long time to cover major distances and this allows navigational errors to accumulate. However, if developing inaccuracy can be reliably corrected en route, then the cruise missile's relatively low speed becomes an advantage. As the cruise missile project manager once put it, "you can always hit the bullseye if you can walk slowly over to the wall and put your finger on it."[11]

The U.S. long-range cruise missiles achieve remarkable accuracy by being able, periodically, to recognize their whereabouts and determine the course changes necessary to "hit the bullseye." This system is known as TERCOM (terrain contour matching). Portions of the selected flight path to the target are surveyed to determine variations in ground elevation, a task done by satellites for many years. These surveyed areas are divided into a matrix of squares. Each square is given a number representing the average elevation of the ground. The resulting digital contour map is stored in the memory of a small computer installed in the missile. When the missile is en route to its target, a radar altimeter starts taking readings before it expects to overfly the surveyed area and stops taking them at an equal distance after it has left that area. The computer compares the information on ground elevation provided by the radar altimeter with the map in its memory. Then it determines how the missile's actual direction of flight differs from that necessary to bring it accurately to the next TERCOM checkpoint and, eventually, to the target. The computer will then issue instructions to bring about the appropriate changes in the missile's direction of flight.

The computer used in both ALCM and SLCM can store contour maps for up to 20 segments of the route to the target.[12] This permits considerable flexibility in plotting the route: known defenses can be bypassed and terrain features such as mountains can be either avoided or exploited to conceal the weapon from enemy radars. Early in 1974, it was stated that on a typical mission a TERCOM check every 200 n.m. would be adequate to achieve the accuracy objective for the ALCM.[13]

In between TERCOM checkpoints, and particularly from launch to the first landfall, the cruise missile is maintained on a set course by means of an inertial guidance system. All inertial systems drift over time; the drift rate for the system in the cruise missile is estimated at between 750 and 900 meters per hour.[14] This error is cumulative so that if the missile flew 1,500 miles at 500 mph, purely on inertial guidance it could be anything up to three kilometers off target at the end. The TERCOM system, in addition to correcting for the navigational errors arising from this drift factor, resets the inertial system so that the error does not accumulate.

The radar altimeter also enables the cruise missile "to fly as low as 20 metres over water, 50 metres over moderately hilly terrain and 100 metres over mountains."[15] This low altitude capability supplements the weapon's small radar cross section to provide a high probability of survival during penetration to the target.

This guidance package is essentially common to both ALCM and SLCM.[16] One notable difference is that the ALCM will carry a bulk storage tape containing TERCOM information on up to ten targets.[17] In the case of the SLCM, because of the large computer capacity available on a submarine, there is no similar storage tape and the computer in the missile is left blank. The computer facilities on a submarine can store TERCOM data on the entire target spectrum and transfer data on the selected target to the missile just before launch. This step is not time-consuming, but the complete process of readying a SLCM for launch takes about 25 minutes. The tactical antishipping version takes about 11 minutes to prepare.[18]

The TERCOM core of the cruise missile guidance system weighs less than 90 lbs. and the full electronics suite (for the SLCM) weighs just 120 lbs. The requirement for the SLCM was that the electronic systems, prior to packaging in the missile, should be a maximum of three cubic feet in volume.

The accuracy potential of the long-range cruise missile is primarily dependent on two factors: the size of the squares in the TERCOM matrices and the distance of the last TERCOM checkpoint from the target. Clearly, if the squares of the matrix are 100 meters on a side, the best the TERCOM system can do is to establish that it is within 100 meters of the flight path that will produce an impact on the target. It is well within the state of the art to do much better than this. Tsipis and others suggest that the squares can be reduced to ten meters on a side. The resolution of available radar altimeters is more than adequate to handle this reduction in size.[19]

In practice, it seems likely that the TERCOM maps are made progressively finer as the missile approaches the target. The first map, on or just behind a coastline, would probably cover a large area with the matrix made up of relatively large squares in order to insure that the missile flies over this area after its inertially guided flight from the submarine or aircraft launch platform. By making the first map relatively large one can allow for inertial drift; for errors in the position of the launch platform arising from inherent limitations in the platform's navigation equipment; and for the possibility that enemy defenses may make it necessary to launch the missile from a considerable standoff distance.

The second major source of inaccuracy stems from the fact that the last TERCOM checkpoint may be some distance from the target owing to the lack of sufficiently uneven terrain in the immediate vicinity of the target. Thus if the inertial drift is 750 meters per hour, the missile is flying at 500 mph, and the last checkpoint is 60 miles from the target, then the missile will impact some 295 feet from the target. If the last checkpoint is 10 miles from the target, this source of inaccuracy is reduced to just 50 feet.

In sum, the operational accuracy of a TAINS-guided long-range cruise missile probably falls in the range 100 to 300 feet. Even the upper figure in this range represents about a threefold improvement in accuracy over the Minuteman III ICBM, currently the most accurate missile in the U.S. strategic arsenal. Coupled with a 250-KT warhead, this accuracy gives the cruise missile single shot kill probabilities that are both high and substantially insensitive to the hardness of the target. Nevertheless, a third component to the cruise missile guidance package is under consideration. This is an optical terminal homing device called scene-matching area correlator (SMAC). SMAC would begin to function after the missile has left its last TERCOM checkpoint. An image of the target is stored in the missile and the weapon will maneuver in the final stages of flight until the observed scene exactly matches the stored image. This requirement for pinpoint accuracy is related primarily to the intention eventually to use the GLCM with a conventional rather than a nuclear warhead for some missions.[20]

The nuclear warhead for both ALCM and SLCM is designated W80. The warhead for the GLCM is basically the same but with sufficient modifications—probably an enhanced radiation feature and perhaps a wider range of yields—to warrant the new designation W84. Development of the warhead started in April 1976 and was expected to take three years.[21] The development period is relatively short because the W80 is not a new warhead but a variant of the W69 device developed in the late 1960s for the SRAM. An important consideration driving the selection of an existing warhead was the Threshold Test Ban Treaty signed in July 1974. This treaty prohibited, as of 31 March 1976, underground nuclear weapon tests having a yield in excess of 150 KT. The treaty had not been ratified by this date, but both the United States and the Soviet Union stated they would observe the limitations during the entire preratification period.

An existing warhead can be refined and redeployed with adequate confidence without testing it at its full yield. The refinements to the

W69 device include an increase in yield, a reduction in weight, and additional safety features.[22] In particular, the high explosives used to produce a critical mass in the nuclear material will be insensitive to shock and heat in order to reduce the risk of accidental detonation. The arming and fusing devices for the warhead will be linked to the completion of the program of navigational checks provided for the missile prior to launch. Thus any missile that fails to complete its assigned TERCOM checks will remain inert.[23]

All open sources put the yield of the W80 at either 200 or 250 KT. Both may be correct since the warhead will have two yields.[24] It is more likely, however, that the alternative yield will be substantially lower than either of these figures since the comment on two yields being available was made in the context of missions requiring very low collateral damage.

As has been seen, the dimensional constraints imposed for cruise missile components were very severe. In the case of the ALCM, at least in its earlier versions with the duckbill nose, the warhead envelope was an irregular shape 36.6 inches long with a maximum diameter of 10.6 inches at the front and 12 inches at the back. The volume of the warhead was 2.34 cubic feet and the weight just 240 lbs. For the SLCM, the warhead envelope is the frustum of a cone 41 inches long with a small diameter of 11.5 inches and a large diameter of 15.25 inches.

One of the strongest arguments advanced in favor of strategic cruise missiles has been their relatively low cost. This argument has two facets: first, relative to other strategic weapon programs, cruise missiles are cheap both to develop and to manufacture; second, the cost of defending against a cruise missile threat is several times greater than the cost of mounting such a threat.

The rapid and unbroken upward trend in the cost of weapons over the postwar period has been well documented.[25] In the case of strategic ballistic missiles, the historical experience in the United States has been that the cost per missile has increased at an average annual rate of 4.8 percent in real terms, which means that the cost doubles in something less than 14 years.[26] A significant downward discontinuity in this trend—a claim that might be made for the long-range cruise missile—would be a development of great consequence, particularly if the cruise missile was regarded as an alternative to the ballistic missile.

The proper context in which cruise missile costs should be judged is that, while circumstances may not have demanded a further diversification of the strategic forces, such a diversification was attractive if it could be secured at a relatively low cost. Thus while other new strategic systems provided the benchmark for assessing what is a relatively cheap system, the cruise missile has never been regarded as a direct competitor with any of these systems. The cancellation of the B-1 in favor of a force of B-52s equipped with long-range ALCMs is a partial exception, but it must be pointed out that this was a competition within the air-breathing component of the strategic Triad, not a competition between its respective components. If cost were the only criterion, the U.S. strategic arsenal would probably consist only of fixed land-based ICBMs. The whole issue concerns the additional assurance provided by different platforms with different survival characteristics launching weapons from different directions and on different trajectories. The bomber force is the most costly leg of the Triad to operate and maintain, but it is retained partly because of its contribution to diversity and, more importantly, because it is still the only strategic system that can be launched without being irrevocably committed. In short, although cost is an important consideration in selecting new strategic weapons, it is far from being the predominant one.

During the 1970s the United States pursued programs to modernize all three legs of the Triad: the B-1 bomber, the Trident SLBM system, and the MX ICBM. At the time of its cancellation, the program acquisition cost of the B-1 was $21.8 billion. The corresponding estimate for Trident SSBNs (as of March 1977) was $21.4 billion; estimates for 200 land-mobile MX ICBMs were of the order of $25 billion. Against these figures, the strategic cruise missile program—and only one strategic variant, the ALCM, remains—is certainly inexpensive if the comparison is made in gross terms. It is quite true that, if the comparison were made in terms of cost per warhead on target, the cost advantage of the cruise missile would shrink and on certain assumptions (particularly regarding vulnerability) might vanish altogether. But, as just pointed out, such a comparison has little validity.

The cost of developing and procuring 3,418 ALCMs as of April 1979 was $4457.1 million. To this must be added the cost of modifying the 150 B-52 carrier aircraft, estimated at $960 million or $6.4 million per aircraft. The total ($5.4 billion) is roughly one quarter that of the other strategic programs cited above. As regards the second dimension of the cost argument, U.S. estimates put the cost of mounting a respectable defense against the cruise missile at about $50 billion.

There are those who maintain that costs for the strategic cruise missile are artificially low because, so far, they are planned for deployment only on existing platforms. The argument goes on that in the not-too-distant future this luxury will no longer be available, and at that time the commitment to cruise missiles will begin to involve expenditures of a similar order of magnitude to the other options for modernizing the strategic forces. The focus of this argument is that the existing fleet of B-52 strategic bombers will become structurally unsound by the early 1990s at the latest, and that before this happens the United States will have to develop and procure a fleet of new cruise missile carriers. In fact, when it cancelled the B-1 bomber in favor of a force of B-52s armed with ALCMs, the Carter administration indicated that it would seriously study the option of augmenting this force with a fleet of wide-bodied aircraft, each carrying in excess of 50 ALCMs. In sum, there was the feeling that the B-52/ALCM force was the thin edge of the wedge as far as strategic cruise missiles were concerned, and that insufficient account was being taken of the cost implications over the longer term of the decision to add these weapons to the strategic arsenal.

Speculating on developments in the strategic forces 15 years or more into the future is a hazardous pastime, particularly if one seeks to identify the cost of alternative developments. A recent study concluded that a fleet of wide-bodied cruise missile carriers would remain the least expensive way of augmenting the strategic forces beyond the improvements already firmly planned.[27] This study, prepared by the Congressional Budget Office, assumed a baseline modernization program that closely followed existing plans: 20 Trident SSBNs with a total of 480 Trident 1 SLBMs, plus 300 B-52s with 3,000 long-range ALCMs. The cost of procuring and maintaining these forces through the year 2000 was put at $120 billion (in constant FY1979 dollars). In the event that additional forces were deemed necessary, the study identified the following four options.

75 wide-bodied aircraft armed with 4,800 ALCMs.
Redeploying 550 Minuteman III ICBMs so that they could be moved about at random between 8,600 shelters, the so-called Multiple-Aim-Point (MAP) deployment mode.
Procuring 200 to 300 MX ICBMs with a MAP deployment system.
Procuring an additional 12 Trident SSBNs and equipping the whole 32-boat force with the Trident 2 SLBM.

The study concluded that the first option would add $15.6 billion to the cost of the baseline modernization program of $120 billion. The closest contender was 200 MX ICBMs, which would add $28.3 billion, while the others fell in the range of $32.2 to $37.6 billion. Subsequent developments, however, point to the likelihood that the United States will acquire a new bomber capable of carrying standoff cruise missiles and of penetrating Soviet airspace to deliver short-range missiles and bombs. The wide-bodied cruise missile carrier is now a rather remote possibility.

NOTES

1. Aviation Week and Space Technology, 18 July 1977, p. 19.
2. Fiscal Year 1977 Authorization for Military Procurement, hearings, Senate Armed Services Committee, March 1976, p. 6,188. [Congressional material has been referenced in abbreviated form. Complete bibliographic information is given in the Appendix.]
3. Fiscal Year 1976 and July-September Transitional Period Authorization for Military Procurement, hearings, Senate Armed Services Committee, April 1975, p. 5,131.
4. Flight International, 12 November 1977, p. 1,411.
5. Electronic Warfare, September-October 1977, p. 55.
6. K. Tsipis, "Cruise Missiles," Scientific American, February 1977, p. 23.
7. Flight International, 1 October 1977, p. 966. An earlier source gives an SFC of 0.1 [Tsipis, "Cruise Missiles," p. 23].
8. Tsipis, "Cruise Missiles," p. 24.
9. Aviation Week and Space Technology, 13 March 1975, p. 13.
10. Aviation Week and Space Technology, 22 November 1976, pp. 14-15.
11. Fiscal Year 1975 Authorization for Military Procurement, hearings, Senate Armed Services Committee, Part 7, April 1974, p. 3,656.
12. Electronic Warfare/Defense Electronics, September-October 1977, p. 55.
13. Fiscal Year 1975 Authorization for Military Procurement, hearings, p. 3,715.
14. The lower figure is from Flight International, 1 October 1977, p. 965, and the upper figure is from Tsipis, "Cruise Missiles," p. 24.
15. Tsipis, "Cruise Missiles," p. 24.
16. The package is usually called TAINS-TERCOM Aided Inertial Navigation System.
17. A single B-52 bomber carrying 20 missiles would therefore represent a potential threat to up to 200 targets.
18. Fiscal Year 1975 Authorization for Military Procurement, hearings, p. 3,654.
19. See Tsipis, "Cruise Missiles," p. 24, and his "Long-range Cruise Missiles," World Armaments and Disarmament, SIPRI Yearbook 1975 (Stockholm: Almqvist and Wiksell, 1974), p. 316. Also Flight International, 1 October 1977, pp. 964-65.
20. Hearings on Military Posture and HR5068, House Armed Services Committee, February-March 1977, p. 1,112. Both the

Senate and House Armed Services Committees have supported a nonnuclear warhead for the GLCM (International Herald Tribune, 23 June 1977, p. 3).

21. Hearings on HR6566 ERDA Authorization Legislation, Intelligence and Military Application of Nuclear Energy Subcommittee of the House Armed Services Committee, February-April 1977, pp. 234, 258.

22. For example, since the W69 was developed, the arming and fusing mechanisms for nuclear warheads have been reduced in weight from 30 to 40 lbs. to about 10 lbs. (ibid., p. 11).

23. Fiscal Year 1975 Authorization for Military Procurement, p. 3,687.

24. Hearings on Military Posture and HR5068, p. 604.

25. See, for example, Norman R. Augustine, "One Plane, One Tank, One Ship: Trend for the Future," Defense Management Journal (April 1975): 34–40. See also Robert L. Perry et al., System Acquisition Experience (Santa Monica, Calif.: Rand Corporation, RM-6072-PR, November 1969), and Ron Huisken, "The Dynamics of World Military Expenditure," World Armaments and Disarmament, SIPRI Yearbook 1974 (Stockholm: Almqvist and Wiksell, 1974), pp. 123-39.

26. Hearings on Military Posture and HR3689 [HR6674], Department of Defense Authorizations for Appropriations for Fiscal Year 1976, House Armed Services Committee, April-May 1975, p. 1,826. In chronological sequence the missiles covered in this computation were: Polaris A-1, Minuteman I, Polaris A-2, Polaris A-3, Minuteman II, Minuteman III, Poseidon C-3, and Trident C-4.

27. Aviation Week and Space Technology, 17 July 1978, p. 22.

2

U. S. LONG-RANGE CRUISE MISSILES: A HISTORICAL NOTE

INTRODUCTION

The purpose of this chapter is to provide some historical perspective on U.S. cruise missiles. Much comment and discussion on ALCM, SLCM, and GLCM gives the impression that the United States has never before seriously considered the cruise missile nor deployed any such weapons on a significant scale. Both impressions are false.

The cruise missile age dawned dramatically. On 13 June 1944 the first of more than 10,000 German V-1s was launched against England.[1] Powered by a pulsejet that provided a speed of some 400 mph out to a maximum range of 160 miles and guided by a preset compass, the one-ton warhead of the V-1 caused considerable damage at first. Subsequently, proximity-fused ammunition for antiaircraft guns and prepositioned fighters were able to bring about an attrition rate close to 95 percent.[2]

The United States was quick to emulate the Germans. A slightly improved version of the V-1, designated the JB-2 and nicknamed Loon, appeared before the end of the war. About 300 examples were built but none, apparently, was used in the war. In addition, the Navy later acquired 349 Loons from the Air Force to investigate their suitability for surface ships, but the project was abandoned around 1951.

The Loon was the first of a number of cruise missile programs that were funded to an advanced stage of development and a few that achieved operational status. Until the early 1950s, the fortunes of the cruise missile, particularly in the medium- and long-range surface-to-surface role, climbed steadily. Then in the space of four or five years they dropped out of the picture almost completely. This transition is well illustrated in observations made in two of the leading journals of the period. In March 1954 the <u>Journal of Space Flight</u> observed that:

16 / ORIGIN OF THE STRATEGIC CRUISE MISSILE

> In the field of long-range ground-to-ground missiles one interesting fact is gradually becoming apparent. Rocket propulsion is losing out to the turbojet.[3]

Four years later <u>Aviation Week</u> stated that, "Few missile men consider air breathers as anything more than stop-gap weapons, existing on borrowed time."[4] A year later <u>Missiles and Rockets</u> observed:

> The surface-to-surface air breathing field has declined as preference for ballistic delivery has grown. Apart from the dark horse of an extremely low level weapon system, most observers predict the decline will continue.[5]

The following discussion of the more important strategic and tactical cruise missile programs illustrates the general accuracy of these observations.

EARLY STRATEGIC CRUISE MISSILES

The missiles that fall into this category include three developed by the Air Force (Snark, Navaho, and Hound Dog) and two by the Navy (Regulus I and Regulus II).

Snark and Navaho

The SM-62 Snark was conceived in January 1946 as a cruise missile capable of traveling intercontinental distances with a high-yield nuclear warhead. At that time nuclear warheads with the requisite yield, presumably of the order of one megaton (MT) and higher, weighed in the range of 4,500 to 6,800 kilograms.[6] The prospect of developing rocket engines capable of lifting these weights over long distances in a ballistic trajectory was considered to be low while aircraft and jet-propulsion technologies were well in hand. A second major stumbling block was the guidance of ballistic rockets. The German V-2, despite its short range (about 320 kilometers), still had a miss distance of 3 to 5 miles.[7] Even a most generous allowance for technological progress could leave one skeptical that at least equivalent accuracy could be achieved, within a reasonable period of time, over distances of more than 8,000 kilometers.

Snark was a conventionally configured pilotless aircraft powered by a single J-57 turbojet and boosted to operational speed directly from its transporter by two JATO rockets. It had a stellar inertial guidance system and weighed about 30 tons at launch. Snark's development program was unduly long. When the first (and only) operational squadron was activated in December 1957 it was already about four years behind schedule. Another year passed before the squadron received its missiles (some 30 all told); the Air Force finally declared it fully operational in the latter half of 1959.

This delay proved almost fatal for Snark. In March-May 1953 the nuclear-test series Operation Castle produced the first solid evidence that thermonuclear weapons would bring dramatic improvement in yield-to-weight ratios. Within two years, after a series of high-level panels, the Atlast intercontinental ballistic missile program was put on a crash basis with top national priority. As the anticipated operational date for Snark and Atlas became increasingly coincident, deployment plans for the former were cut back.[8]

Nevertheless, for a short time Snark was the only operational U.S. intercontinental missile, threatening the Soviet Union with a thermonuclear warhead with a reported yield of 4 MT. Moreover they were hectic times. The imminent availability of Snark was advanced by some as a reason not to be unduly concerned about Sputnik, and its eventual operational deployment (over 18 months after this event) was widely acclaimed.

It is clear, however, that the Snark program was maintained almost solely as a hedge against the increasingly remote contingency of a significant delay in the ICBM program. Although it flew higher (around 60,000 feet), somewhat faster, and had a smaller radar cross section than the B-52, the Snark had a small and unaugmentable ECM (electronic countermeasure) capacity. Even when it was first deployed, U.S. military planners were apparently aware of its vulnerability to their own surface-to-air missiles, and Soviet technology was very strong in this area.[9] The Snark's vulnerability when in flight was apparently more than matched by the difficulty of getting it into the air. In 1976 Vice-Admiral Gerald E. Miller, a former deputy director of the Joint Strategic Target Planning Staff, recalled that the chances of getting the weapon off the launch pad were about one in three.[10] It was primarily on the ground of unreliability that Secretary of Defense McNamara, after a briefing on U.S. strategic war plans soon after he assumed office, took steps to have Snark removed from the inventory. In any event, the Air Force had effectively signalled its abandonment of cruise missiles for the intercontinental strike role back in July 1957 when it canceled the follow-on to the Snark, the Navaho.

The design of Navaho was finalized in 1950; from a technological point of view it was a major step forward from Snark. Whereas Snark was subsonic and had an inertial guidance system supplemented by stellar observations, Navaho cruised at Mach 3 and its guidance system was wholly inertial. Moreover, Navaho was so large (some 300,000 lbs.) that a significant development effort was required to provide the rocket engines that would boost it to supersonic speed to allow the ramjet cruise engines to take over.

The Air Force was quite severely criticized for pursuing Navaho so long. Proponents of ballistic missiles regarded the weapon as having pronounced dinosaurian qualities and considered the $970 million spent on its development as largely wasted. As it turned out, many of the technological advances achieved under the Navaho program found application in subsequent weapon systems. Moreover, the Air Force had to contend with conservative elements in the Congress. Several members of the armed services committees were concerned over the fact that Navaho was canceled and planned production of Snark severely cut back at a time (September 1957) when a successful ICBM test had yet to be conducted.[11] It could still be argued, however, that Navaho demonstrated that the assumptions made in the late 1940s, namely, that propulsion and guidance technologies were more attainable for long-range cruise missiles than for ballistic missiles, were wrong. In pursuing the Navaho the Air Force was ignoring the many advantages of the ballistic missile—including short reaction time, short flight time to the target, and relative east of protected deployment—and the fact that the development of the Navaho itself required solutions to several of the major technical barriers that still stood in the way of long-range ballistic missiles, notably reliable rocket engines and inertial guidance systems. The Air Force should have realized earlier that guiding a missile accurately to a target is inherently more difficult for a cruise missile with a flight time of several hours than for a ballistic missile with a flight time of 30 minutes. In short, in persevering with the Navaho, the Air Force displayed its institutional affinity with the air-breathing missile, a weapon that looked and performed like an aircraft and could be regarded as a complement to the manned bomber.

Hound Dog

The AGM-28A/B Hound Dog was a turbojet-powered, air-to-surface missile carried by the B-52 strategic bomber. Originally intended as an interim weapon, pending the availability of the Skybolt air-launched, intermediate-range ballistic middle, Hound Dog remained operational for 16 years until withdrawn from service in 1976.

Hound Dog was a significant beneficiary of the technologies developed for Navaho, particularly the guidance and airframe design.[12] North American, the contractor for both missiles, reported that a substantial portion of the personnel engaged on Navaho were transferred to the weapon system 131B project of which Hound Dog was the missile component.[13] As deployed, Hound Dog was over 13 meters long, weighed approximately 4,500 kg., and cruised at Mach 1.6 over a maximum range of about 800 kilometers. A B-52 could carry two missiles, one under each wing. Judging from inventory levels (estimated at about 600 missiles in 1969),[14] several hundred aircraft were configured to carry the weapon.

The Strategic Air Command (SAC) was enthusiastic about Hound Dog. General Powers, commander in chief of SAC at the time Hound Dog was first deployed, described it as "a tremendous weapons system."[15] In a speech to the Economic Club of New York in January 1960, he elaborated as follows:

> Its primary significance lies in the fact that it will vastly increase the utility and flexibility of the manned bomber and permit a variety of new tactics, such as attacks on additional targets in different areas at the same time.[16]

Not surprisingly, this statement closely resembles the rationales advanced more than a decade later for the ALCM. It seems that even in the late 1950s there was concern about the vulnerability of strategic bombers penetrating Soviet airspace at high altitude. Alain Enthoven, the head of the system analysis division in the Office of the Secretary of Defense (OSD) from 1961 to 1968, reported that SAC changed its B-52 tactics to low-level penetration as early as 1957.[17] Early in 1958 Secretary of Defense McElroy made the forecast that Hound Dog would be only the first of a series of air-launched missiles designed to preserve the viability of the penetrating strategic bomber. To quote him:

> We are going to do more of this kind of thing to continue to make the bomber, the manned bomber, perform its very special kind of mission in relation to all the other combined strategic delivery systems which together make a far more effective deterrent threat than if we confine our effort to some single delivery system.[18]

Regarding the Hound Dog's performance the public record is rather thin. There is some evidence that the weapon's accuracy was less than adequate. A former Air Force officer interviewed for this study claimed it was never regarded as a primary strike weapon due to its inaccuracy: a CEP (circular error probable) at full range in excess of one mile. On the other hand it carried a very large thermonuclear warhead. Occasional references refer to its yield as "in the megaton range," and for several years one standard source referred to a yield of 4 MT.[19] The designation AGM-28B was given to a version with improved navigational accuracy. Of greater interest in the present context was the attempt in the 1960s to give Hound Dog a TERCOM capability, but this project was abandoned.[20]

The final attempt to upgrade Hound Dog was a project called Hound Dog II. This was a highly classified program, but it seems fairly clear it was designed to give Hound Dog the ability to attack Soviet airborne warning and control aircraft, in addition to attacking targets on land.[21]

It is apparent that the Hound Dog constitutes a thread of continuity in Air Force interest (or at least Department of Defense interest) in bomber-launched strategic cruise missiles. No similar continuity of interest was evident in the Navy.

Regulus

The SSM-8A Regulus I was a turbojet-powered Navy missile primarily intended to be launched from submarines against fixed targets on land. It was in operational service from 1955 to 1966, and at various times was deployed on aircraft carriers and cruisers as well as submarines. Regulus I carried a nuclear warhead at high subsonic speed to a maximum range of 440 nautical miles. Having a low flight profile, it was regarded as difficult to defend against. On the other hand, the fact that the submarine had to surface to enable the weapon to be pulled from its storage hangar and fired was regarded as a serious deficiency. The Navy did not press to expand its deployment on such platforms beyond a peak force of five boats.[22]

Nevertheless, the Navy pursued a successor missile, the XSSM-9 Regulus II. Regulus II had the same launch sequence as Regulus I but had more than double the range (1,000 n.m.), a cruise speed of Mach 2, and an all-inertial guidance system. Viewed as a strategic weapon, Regulus II was no match for the Polaris SLBM. Although Polaris was originally scheduled to become operational in 1963, the combination of an uncommonly smooth development program and technical compromises (particularly on range) advanced its operational date to mid-1960.[23] At best Regulus II would have bettered this date by a few months. The Navy canceled the program in December 1958 after an investment of some $290 million.

Although the Navy acknowledged that Regulus II had a high unit cost and, when compared with Polaris, was technologically obsolete, there was some reluctance to cancel the program.[24] As far as one can judge from the public record, the debate, to some extent within the Navy but mostly between the Navy and OSD, revolved on the issue of whether Regulus should be viewed exclusively as a strategic weapon. Early in 1958 Rear Admiral Clark, director of the guided missile division of the Office of the Chief of Naval Operations, expressed the view that Regulus II would be a more flexible weapon than Polaris. Its range and payload (both in terms of weight and type) were variable, making it a viable weapon in situations short of all-out nuclear war.[25]

Years later when the current cruise missile programs were proposed, several strands of explanations emerged as to why the United States did not pursue cruise missiles for tactical roles while the Soviet Union fielded an increasingly diversified family of such weapons. In congressional testimony in 1971, John Foster, the director of defense research and engineering (DDR&E), advanced the argument that the United States did not divert the naval cruise missile program to tactical roles because potential enemies provided no targets.[26] This was a very plausible explanation. Given the size of the Soviet surface fleet at that time and its predominant coastal-defense orientation, it would have been difficult to justify a tactical naval cruise missile, particularly in view of the massive offensive capability available from carrier-borne aircraft. Foster was presumably summarizing the OSD counterargument to Navy pressure to save the Regulus program by giving it a tactical rather than a strategic orientation.

Navy officials offered a more humble explanation. In March 1974 the chief of naval operations, Admiral Zumwalt, told a House Appropriations Subcommittee that the Navy did not foresee the great reduction in its carrier force that in the late 1950s was considered to be an effective counter to Soviet naval cruise missiles and their platforms.[27] Earlier, Vice-Admiral Semmes, deputy CNO, had put the matter more bluntly: "We were not smart enough to move that missile [Regulus II] into a tactical application and we should have."[28]

The issue of why the Navy thereafter displayed no visible interest in cruise missiles will be taken up again in the next chapter. For the moment one can note that the Navy was not developing any cruise missiles after Regulus II was canceled in December 1958, and not producing any cruise missiles after January 1959 when the last of 514 Regulus I missiles was delivered.[29]

EARLY TACTICAL CRUISE MISSILES

The focus of this section will be on two Air Force weapons, Matador and Mace. Both were medium-range missiles with optional nuclear, conventional, or chemical warheads, but primarily intended and deployed as tactical nuclear weapon delivery systems.

The TM-61A Matador became the Air Force's first operational guided missile when it was deployed in 1954 and, together with the Army's Honest John artillery rocket and Corporal missile, constituted the vanguard of the tactical nuclear weapon systems sent to Europe. The First Pilotless Bomber Squadron (light) was activated in October 1951, but it took another three years for the Air Force to become sufficiently familiar with the new weapon to deploy it to Europe.

Development of the Matador began in 1946. Built around a widely used jet engine, it closely resembled and was not significantly smaller than some of the fighter aircraft of that time.[30] It was, however, mobile and featured a zero-length launch, being propelled to flight speed by a rocket booster.

The weapon's one devastating weakness was its guidance system. The MSQ-1 command control guidance system required that radar contact be maintained with the weapon throughout its flight. This meant that: Matador's effective range was considerably less than its potential range (reportedly about 1,000 km); it was highly vulnerable to enemy electronic countermeasures; and the number of missiles that could be en route to their targets at any one time was severely limited.[31]

Each of these drawbacks was at least partially alleviated in the TM-61C model, which became operational in 1954. By the time Matador production ended in 1957, some 1,200 TM-61C models had been produced at an average unit cost of about $60,000.[32] A statistical analysis of 78 scheduled training launches of the TM-61C over the period April 1957 to September 1960 yielded the following data:[33]

(a) launch reliability, 95 percent; in flight reliability, 75 percent; overall reliability 70 percent.
(b) probability of launching any number of missiles without a delay during the launch sequence, 59 percent.
(c) average CEP 2,700 ft. However, this varied enormously with the type of crew; instructors achieved an average CEP of 1,600 ft. while student crews could only manage 3,230 ft.
(d) during the "Operation Marblehead" exercises conducted in Tripoli, June-October 1958, the TM-61C CEP was 1,980 ft.

The history of the TM-61B is somewhat more involved. In 1949 a missile guidance concept known as ATRAN (automatic terrain recognition and navigation) was developed, and in 1952 the Air Force decided to incorporate it in a version of the Matador. The ATRAN system consisted of a search radar, a map-matching device, and a terrain clearance controller. During flight the map-matching device compared the images provided by the search radar with a radar photograph of the terrain overflown that was inserted prior to launch. Errors between the two images were broken down into longitudinal and lateral

AMERICAN LONG-RANGE CRUISE MISSILE / 23

components and the missiles' course adjusted accordingly. The map stored in the missile's memory was at first a direct radar image, but subsequently it became possible to fabricate the maps synthetically from topographical charts. Development of the ATRAN system proved extremely difficult, and the schedule for the "B" model of the Matador slipped well beyond that of the "C." A series of tests in 1955-56 produced poor results. ATRAN demonstrated an ability to guide a modified TM-61A in a straight line and through predetermined turns, but only out to ranges of about 40 miles before getting lost. It was determined that the most probable reason was the poor quality of the radar maps.[34] The Air Force persevered, however, and the new weapon was deployed to Europe in 1959 after having been redesignated as the TM-76A Mace.

Slightly longer and heavier than the Matador, the Mace A was widely acclaimed. During tests ATRAN had demonstrated a strong immunity to countermeasures and it could be programmed to penetrate at altitudes as low as 500 ft.[35] The commander of the first Mace squadron in Europe described it as "capable of penetrating any known air defense."[36] Flight altitude was variable and in a high-low approach Mace A could attack targets out to about 1,200 km. A rapid-fire multiple launch system was installed in mid-1961, which made it possible to launch four missiles in less time than it formerly took to fire one. Each tactical missile group (three groups made up the 38th Tactical Missile Wing based in the Federal Republic of Germany) was equipped with 36 missiles, of which 24 would be on alert at any one time with 6 capable of being launched within 12 minutes and all 24 within 16 minutes.

There was a substantial cost penalty associated with this new capability. Whereas the TM-61C had a unit cost of $60,000, the TM-76As were expected to cost $250,000 each at peak production rates. Moreover, very little of this fourfold increase would have been due to inflation.

A second and last version of Mace, the TM-76B, was deployed in 1961. Designed primarily for high altitude penetration, it had an all-inertial guidance system and a range of about 2,200 km.

The role of both Matador and Mace, at least in Europe, was to complement the fighter-bomber aircraft designated to deliver tactical nuclear weapons. Presumably, the missiles were intended to overcome the limitations of manned aircraft in darkness and adverse weather, and to attack targets in areas of enemy air superiority. For some reason (or reasons) neither Matador nor Mace was a particularly visible weapon in terms of congressional interest, treatment in the more prominent trade journals, or in assessments of the military balance during the period of their deployment. This is rather surprising since they were nuclear systems and for 15 years after they were

24 / ORIGIN OF THE STRATEGIC CRUISE MISSILE

first deployed they were by far the longest-range tactical nuclear missile systems in the U.S. arsenal.[37]

An exception, as far as congressional interest is concerned, was an attempt to cancel the Mace B program. In fall 1958 when the fiscal year 1960 defense budget was being drawn up, the Bureau of the Budget questioned the need for the long-range Mace B in view of the two IRBM programs, the Air Force Thor and the Army Jupiter, that were nearing completion. The House apparently accepted this argument and denied the $127 million requested for the Mace B, but the program survived the House-Senate Conference Committee after the secretary of defense appealed against this decision.[38]

Mace B remained in service for nearly a decade. In February 1968 Secretary of Defense McNamara announced that 18 of the 96 Mace Bs in the Federal Republic of Germany would be withdrawn during FY1969 as the Army's Pershing 1A (maximum range 840 km.) took over their Quick Reaction Alert (QRA) role.[39] He also stated that the rest of the Mace B force would remain operational, "at least for the next few years."[40] One source suggests that they were finally withdrawn from Europe in 1970,[41] but they survived beyond this date in South Korea and Taiwan.

In sum, the Matador/Mace series provided an additional thread of continuity in Air Force interest in cruise missiles. Although active development of cruise missiles ended with the deployment of Mace B in 1961, sustained Air Force interest in and the basic viability of the weapon could be said to be indicated by the fact that Army missiles and Air Force fighter-bombers provided constant competition for the QRA role. In the case of the Navy, the turn-away from cruise missiles was rather more decisive. Given the Navy's open acknowledgment that Regulus I was a seriously flawed weapon, it probably remained in service as long as it did (1955-66) more because it was there and because three submarines had been built solely to carry it than for any other reason.

NOTES

1. Basil Collier, The Battle of the V-Weapons: 1944-1945 (New York: William Morrow, 1965), pp. 163, 180.
2. Nels A. Parson, Jr., Missiles and the Revolution in Warfare (Cambridge, Mass.: Harvard University Press, 1962), pp. 26-27. As an interesting aside in view of what will follow in Ch. 4, it is worth recording the view that the V-1, a Luftwaffe weapon, came into being because the Air Force was envious of the Army's large program at Peenemunde working on what became the V-2 ballistic rocket.
3. Journal of Space Flight, March 1954, p. 2.

4. Aviation Week, 3 March 1958, p. 3.
5. Missiles and Rockets, 2 February 1959, p. 11.
6. The weapon dropped on Hiroshima weighed about 4,000 kilograms and its explosive yield was estimated at 12.5 kilotons.
7. Werner Von Braun and Frederick I. Ordway, III, History of Rocket and Space Travel (London: Thomas Nelson and Sons, 1967), p. 121.
8. Several in-depth studies of the development of the ICBM have appeared. The most recent is Edmund Beard's Developing the ICBM (New York: Columbia University Press, 1976).
9. Investigation of National Defense Missiles, hearings before the Committee on Armed Services, House of Representatives, January-February 1958, p. 4,769.
10. First Use of Nuclear Weapons: Preserving Responsible Control, hearings, Subcommittee on International Security and Scientific Affairs of the Committee on International Relations, House of Representatives, March 1976, p. 66.
11. See, for example, Enquiry into Satellite and Missile Programs, hearings before the Preparedness Investigating Subcommittee of the Committee on Armed Services, Senate, November-December 1957 and January 1958, p. 867.
12. Von Braun and Ordway, History of Rocket and Space Travel, p. 122.
13. Enquiry into Satellite and Missile Programs, p. 2,238.
14. R. T. Pretty and D. H. R. Archer, eds., Jane's Weapon Systems 1970-71 (London: Sampson, Low, Marston, 1970), p. 103.
15. Missile and Space Activities, joint hearings before the Preparedness Investigating Subcommittee of the Committee on Armed Services and the Committee on Aeronautical and Space Science, Senate, January 1959, p. 130.
16. Speech reproduced in Missile and Space Activities, hearings, p. 6.
17. Status of US Strategic Power, hearings before the Preparedness Investigating Subcommittee of the Committee on Armed Services, Senate, April 1968, p. 214.
18. Enquiry into Satellite and Missile Programs, hearings, p. 2,477.
19. John W. R. Taylor, ed., Jane's All the World's Aircraft 1960/61 (London: Sampson, Low, Marston, 1960), p. 464.
20. Jane's Weapon Systems 1970/71, p. 103.
21. Fiscal Year 1972 Authorization for Military Procurement, hearings, Senate Armed Services Committee, March-May 1971, p. 3,083. A similar dual-purpose capability is now proposed for the advanced strategic air-launched missile (ASALM).

22. *Enquiry into Satellite and Missile Programs*, hearings, pp. 748, 1,730. Of the five, three were purpose-built to carry four or five Regulus missiles each. The other two were converted attack submarines with two missiles each.

23. The first Polaris patrol began in November 1960, but the weapon's operational readiness was demonstrated in July 1960 when the George Washington launched two missiles while submerged.

24. *Major Defense Matters*, hearings before the Preparedness Investigating Subcommittee of the Committee on Armed Services, Senate, March 1959, p. 187.

25. *Investigation of National Defense Missiles*, hearings, p. 4,695.

26. *Fiscal Year 1972 Authorization for Military Procurement*, hearings, p. 514.

27. Related in *Space Business Daily*, Washington, D.C., 11 March 1974, p. 50. This may have been a polite way of rebuking the Congress for not funding more new carriers.

28. *Fiscal Year 1972 Authorization for Military Procurement*, hearings, p. 981.

29. Two other Navy cruise missile projects can be mentioned in passing: the Rigel, a Mach 2, submarine-launched ramjet-powered weapon canceled in 1952 in favor of Regulus I; and the XSSM-N-2 Triton, an advanced long-range (1,500 n.m.), high speed (Mach 3.5), submarine-launched weapon canceled in 1956 in favor of Polaris.

30. For example, the F-84 Thunderjet.

31. The guidance system was not the only factor limiting launch frequency. Preflight checkout requirements on the early Matadors were so cumbersome that in 1953 a demonstration of *maximum* effort involved the launching of five missiles in a 24-hour period. ("History of the First Pilotless Bomber Squadron (light), 1 January 1954-1 June 1954," document read at the Albert F. Simpson Historical Research Center, Maxwell Air Force Base, Alabama.)

32. Martin Company press release, 22 April 1957. A higher figure ($90,000) is given in *Missiles and Rockets*, February 1958, p. 82. These figures exclude the cost of the nuclear warhead. The last Matador was phased out during FY1962.

33. "TM-61C Operational Summary," document read at the Albert F. Simpson Historical Research Center, Maxwell Air Force Base, Alabama.

34. "Test Report on the TM-76A," document read at the Albert F. Simpson Historical Research Center, Maxwell Air Force Base, Alabama.

35. Ibid.

36. Quoted in Martin Company press release, 7 August 1959. Even if true this capability was soon negated. According to *Jane's*

Weapon Systems 1969-70 (p. 1), Mace A was withdrawn from service before Mace B because the ATRAN system became vulnerable to ECM.

37. The various Army missiles, their date of introduction, and maximum range, were as follows: Corporal, 1954, 140 km.; Redstone, 1957, 320 km.; Sergeant, 1962, 135 km.; Pershing 1, 1963, 480 km.

38. Major Defense Matters, hearings, p. 313.

39. The Mace apparently had a somewhat sensitive mission. In April 1969 a senator asked Air Force Chief of Staff McConnell why Mace was being withdrawn from Europe and what was replacing it. McConnell's entire response was deleted from the public record (Authorization for Military Procurement, Research and Development, Fiscal Year 1970 and Reserve Strength, hearings, Senate Armed Services Committee, March-April 1969, p. 1,000).

40. Authorization for Military Procurement, Research and Development, Fiscal Year 1969, and Reserve Strength, hearings, Senate Armed Services Committee, February-March 1968, p. 213.

41. Norman Polmar, Strategic Weapons: An Introduction (New York: Crane, Russak, 1975), p. 150.

3

THE SUBMARINE-LAUNCHED CRUISE MISSILE: A WEAPON IN SEARCH OF A MISSION

INTRODUCTION

As seen in the previous chapter, the United States Navy essentially abandoned cruise missiles in 1958 with the cancellation of Regulus II. An unsuccessful effort was made to retain a surface-to-surface cruise missile program by stressing its utility for tactical roles, while conceding that the Polaris SLBM had rendered it obsolete as a strategic weapon. The main reason given for not accepting this reorientation was that, given the size of the Soviet surface navy, the United States already possessed a surfeit of tactical offensive capability in its carrier-borne aircraft.

The cogency of this argument was beyond dispute. In 1960 the Soviet Navy possessed some 260 surface units displacing 1,000 tons or more, and approximately one third of these displaced less than 2,000 tons.[1] At that time the United States had operational about 775 major surface warships including 25 attack and support aircraft carriers and about 30 other carrier units configured for antisubmarine warfare, amphibious landing, and escort roles. The aircraft capacity of the 25 attack and escort units was between 1,150 and 1,450, of which some two thirds consisted of fixed-wing fighters, fighter-bombers, and attack aircraft. This was a formidable force with long-range striking power augmented, from 1959 onward, by the AGM-12 Bullpup family of air-to-surface missiles. The main strength of the Soviet Navy lay in submarines—242 ocean-going boats and over 120 coastal boats in 1960[2]—but, in the aggregate, the naval balance was overwhelmingly in favor of the United States.[3]

During the 1960s the Soviet Navy was transformed in two ways. On the one hand, while the number of major surface ships remained roughly constant, a gradually increasing proportion consisted of larger, longer-range, heavily armed units. On the other hand, deployment patterns changed significantly. A permanent squadron appeared in the Mediterranean in 1964 and in the Indian Ocean in 1968, while the Northern and Pacific fleets began to range wider and for longer periods over the Atlantic and Pacific Oceans. Toward the end of the 1960s it was becoming apparent that U.S. naval supremacy was no longer unchallenged.

In U.S. naval circles, criticism began to focus on the concentration of offensive surface-to-surface capabilities on the shrinking force of aircraft carriers. The extrapolation of the World War II aircraft carrier concept began to yield results that were so expensive— to build, to equip with aircraft, and to operate and maintain—that the number of units the United States could afford to keep operational fell almost continuously. The U.S. Navy currently operates 13 carriers. Even this force has been under constant pressure, although the events in Iran and Afghanistan will probably stave off further reductions for a time. Moreover, the Navy itself remains divided on the efficacy of a large, multipurpose aircraft carrier.

A small impetus to diversify the Navy's surface-to-surface capability beyond carrier-borne aircraft emerged as early as the Cuban missile crisis in 1962. The Cuban navy possessed a handful of Soviet-supplied Komar patrol boats armed with the SSN-2 Styx antiship missile, which had a range of 42 kilometers. The lack of a U.S. counterweapon of comparable range, particularly at night or in adverse weather when aircraft performance was seriously degraded, led to the requirement that naval surface-to-air missiles be given some surface-to-surface capability. Accordingly, the RIM-66/67 Standard, successor to the Terrier, Talos, Tartar family of naval surface-to-air missiles, was developed with this additional capability in mind.[4] However, the Standard missile was not deployed until 1969.

A far more significant catalyst to U.S. Navy interest in cruise missiles was the 1967 sinking of an Israeli destroyer by Egyptian-owned SSN-2 Styx missiles. The Navy's first reaction in February 1968 was to establish an Antiship Missile Defense office, which has since presided over a growing list of programs including Aegis, Vulcan-Phalanx, and the basic point missile defense system. In addition, and apparently with some prodding from the Office of the Secretary of Defense, the Navy requested a small amount of money for FY1969 to study air- and ship-launched antiship missiles.

The antiship cruise missile was the subject of conceptual studies for the next two-and-one-half years, culminating in a Defense Select Acquisition Review Council (DSARC) meeting in November 1970 that approved the initiation of development of the AGM-84A (air-launched) and RGM-84A-1 (ship-launched) missiles. The popular name given to these missiles was Harpoon. In itself, the Harpoon is not particularly germane to this study, but it is intimately linked with the long-range cruise missile. Harpoon was designed for a maximum range of 60 nautical miles, with cruise propulsion provided by a tiny turbojet weighing just over 100 lbs. but providing 650 lbs. of thrust. The missile is fully autonomous after launch and its 500-lb. warhead is capable of sinking destroyer-size vessels.

Early in 1971 the Navy announced it was planning a third version of the Harpoon to be launched from attack submarines.[5] This version was designated UUM-84 "encapsulated" Harpoon.[6]

Coincident with the encapsulated Harpoon, the Navy launched another new program called the advanced cruise missile. This program survived for barely two years. It consisted predominantly of conceptual studies and was highly classified. It is rather difficult to piece together the details, but since it constitutes the intervening link between Harpoon and SLCM the attempt is worthwhile. Before doing this, however, it is interesting to speculate on why the Navy began to look beyond the Harpoon at such an early stage. It is standard practice to initiate the development of a successor weapon well before its predecessor goes into operation. On the other hand, two or three years later, the Navy acknowledged that it had no urgent requirement for a successor to the Harpoon and that it was pursuing a tactical variant of the SLCM merely because it was a bargain in the sense that the strategic program would support the bulk of the development costs.

Several factors probably contributed to the initiation of the advanced cruise missile project. One of these was a major study conducted in 1970 by the Center for Naval Analyses that concluded favorably on the efficacy of long-range cruise missiles.[7] Another may have been the unsuccessful attempt by the United States to bring the issue of Soviet cruise missiles into the SALT negotiations.[8] Finally, the Air Force subsonic cruise armed decoy, in pursuing its program, had proven that small turbofan engine technology had reached an advanced stage. One could probably add that the Navy was not anxious to permit the Air Force to monopolize the field of long-range cruise missiles.

The advanced cruise missile, for which funding was first requested in the FY1972 budget (that is, the budget submitted to Congress early in 1971), was to be a tactical antiship missile with a range of about 300 miles. It was also to be an entirely new weapon, a growth version of the Harpoon having been rejected in studies conducted during FY1971.[9] The initial concept was a weapon that would be launched from vertical tubes in a new, purpose-built class of submarines with 20 tubes per boat.[10] This concept significantly reduced the constraint on size and the missile was relatively large, about 34 inches in diameter. Sometime during 1971, John Foster, then director of defense research and engineering, requested the Navy to examine the possibility of achieving ranges of strategic significance with cruise missiles.

The advanced cruise missile then went through various iterations and emerged as a weapon 28 inches in diameter to be fired from horizontal tubes.[11] This concept still implied a new class of submarine or costly modifications to existing submarines.

The FY1973 budget, presented early in 1972, contained a modest request for the advanced cruise missile. At the same time a $2 million supplementary request for FY1972 was submitted for the strategic variant on the ground that the concept was sufficiently attractive to be worth pursuing faster than the approval of FY1973 funding would allow. The assistant secretary of the Navy, in a statement somewhat out of proportion to a request for $2 million, claimed that "the same rationale and sense of urgency that forced us to accelerate ULMS [the undersea long-range missile system, subsequently called Trident], forces us to proceed as expeditiously as possible with advance cruise missile development, for the applications mentioned."[12] As shall be seen below, the acceleration of the Trident SLBM program was based in part on a rather alarming assessment of how far the Soviet buildup of its strategic forces would go. Another reason, alluded to by Defense Secretary Melvin Laird early in 1973, was to help the United States drive a hard bargain with the Soviet Union on the limits on SLBMs to be set in the interim SALT agreement on offensive weapons.[13]

The advanced cruise missile program was canceled by the Navy in November 1972 in favor of a tactical variant of the SLCM, or strategic cruise missile as it was first called. At the time of cancellation only $10.6 million had been spent, but the work did go as far as the testing of models in wind tunnels.

THE STRATEGIC CRUISE MISSILE PROPOSAL

The SALT I agreement in May 1972 prompted a number of amendments to the FY1973 defense budget then under review by Congress. One of these was a request for $20 million to initiate a separate strategic cruise missile (SCM) program. The size, mode of launch, and launch platform for the SCM were not defined, but the following options were under consideration:[14]

Option 1: cruise missiles launched vertically
    from converted SSBNs;
Option 2: cruise missiles launched horizontally
    from SSNs;
Option 3: cruise missiles launched horizontally
    from SSBNs; and
Option 4: cruise missiles launched vertically
    from a new class of SSNs.

The cruise missiles under consideration ranged from 21 inches to 32 inches in diameter and from 246 inches to 345 inches in length, with the smaller designs relating to options 2 and 3 and the larger versions to options 1 and 4. The Polaris A-3 missile carried by the 10 oldest U.S. SSBNs is 54 inches in diameter and just over 375 inches long. Thus deployment on old SSBNs, most of which would have to be retired as the new Trident boats became operational, would permit either a long, large-diameter cruise missile or a cluster of slender missiles to be put into each launch tube.

When he submitted the strategic cruise missile proposal, Defense Secretary Laird expressed a strong preference for option 1, while the Chief of Naval Operations (CNO) Admiral Zumwalt preferred the greater deployment flexibility that would stem from constraining the size of the missile to fit a torpedo tube.[15] As just mentioned, this approach would not preclude deployment of cruise missiles on SSBNs in place of the Polaris SLBM, but this concept gradually faded from the scene.

Accordingly, the Navy pressed for the adoption of a fifth option: to develop a cruise missile with both strategic and tactical applications and compatible with all existing potential launch platforms. This view prevailed and was cemented in the operational requirement document on the sea-launched cruise missile released on 15 November 1974.[16] The SLCM document specified a weapon: sized to fit torpedo tubes 21 inches in diameter and 246 inches long; capable of both strategic and tactical missions with development priority to be given to the former and, for the tactical version, to surface ship platforms; and adaptable to launch from land-based platforms.

The objective for accuracy, deleted from the published document, was apparently a circular error probable of 0.1 nautical miles or about 600 feet.[17] The RDT&E cost of this program, extending over the period FY1973-80, was put at $714.8 million, considerably less than the $918 million figure presented in mid-1972 when the strategic cruise missile was first proposed.[18]

THE OFFICIAL RATIONALE

The proposal to develop and deploy a strategic sea-launched cruise missile was controversial from the outset. For several years the Defense Department in the United States was unable to construct a convincing rationale for this weapon and, in a sense, the effort was eventually abandoned. The program is still very much alive, but persistent skepticism in Congress and in the arms control community, the weapon's prominence in the SALT negotiations, and a developing enthusiasm for it within NATO significantly altered the original concept.

The strategic cruise missile program proposed by Defense Secretary Laird in June 1972 involved a single weapon—a long-range, submarine-launched cruise missile. The platforms envisaged for the SLCM were "a few SSNs" and the ten oldest SSBNs (George Washington and Ethan Allen classes), with Secretary Laird expressing a clear preference for the latter. Laird's justification for proposing SLCM appeared essentially to be a reaction both to the U.S. experience in negotiating the SALT I accords and to the terms of those accords, particularly the interim agreement on offensive weapons.

The reaction, both in the Congress and elsewhere, was generally one of surprise and disappointment. Earlier in 1972, when submitting the FY1973 defense budget to Congress, Laird had delivered a surprisingly strong statement on Soviet intentions:

> It is my belief that the Soviet Union is on a course designed to achieve superiority and that they want to have strategic superiority, naval superiority and conventional ground force superiority.[19]

In general, Laird's budget, particularly in the strategic area, reflected this sentiment and included accelerated development of the Trident SLBM. Far from diminishing this sense of urgency, the signing of the SALT I accords prompted the Defense Department to submit a supplement to the FY1973 budget requesting a start on an entirely new strategic weapon (SLCM) and additional money for five other strategic programs.[20] Further, Laird stated that his support for the SALT agreements was conditional on congressional acceptance of both the strategic program in the FY1973 budget and the special supplemental request.

On the specific issue of SLCM, Laird endeavored to explain the contradiction by arguing that the United States had been able to negotiate effectively in SALT I because it had active development or procurement programs in the various areas of weaponry covered by the agreements, namely ABMs, ICBMs, and SLBMs. During the negotiations the Soviet Union apparently resisted attempts to include their submarine-launched cruise missiles in the interim agreement on offensive weapons, and U.S. officials concluded that their bargaining position on this issue was weak because the United States lacked comparable weapons and did not have any program to develop such weapons.

The fact that cruise missiles were discussed at SALT is clear from the following:
Senator Hughes: Is it correct that the United States proposed and the Soviet Union tentatively accepted, approximately at the time of the third session of the SALT

negotiations, a mutual ban on the development of strategic cruise missiles?

Defense Department: The U.S. and the USSR in SALT I discussed cruise missiles of intercontinental range.[21]

Other military officials pointed rather vaguely to the strategic threat posed by these Soviet weapons, particularly the submarine and ship-launched SSN-3 Shaddock. Further, the existence of the Soviet weapons was used to parry criticisms that the U.S. was starting a "cruise missile race" and, in a similar vein, to argue that any problems these weapons posed for arms control already existed.[22]

Thus one strand of justification was bargaining leverage, to make limitations on cruise missiles more probable in a SALT II agreement. In Laird's words:

I believe one of the important things for us to do is to have the capability to use cruise missiles in those [Polaris] submarines. . . . in view of the fact that there was no limitation placed on this kind of missile system, and the Soviet Union already has close to 70 cruise missile submarines . . . this is a very important program for us to push at this time.[23]

In an interview in February 1979, Henry Kissinger recalled he had similar motives for endorsing the cruise missile proposal, although he couched his argument in broader terms than did Laird.[24] According to Kissinger, U.S. negotiating strategy in SALT I was, of necessity, very much defensive. The Soviet strategic program had a powerful quantitative momentum that the United States would not be in a position to match for several years, that is, until the Trident, B-1, and MX systems had been developed. Adding the cruise missile to this list would provide additional insurance that, when the five-year interim agreement on offensive weapons expired, the United States would be in a position to catch up quantitatively or at least present a credible image of being able to do so.

A second strand of justification was based more exclusively on military grounds. The most enthusiastic proponent of the SLCM in this regard was DDR&E John Foster. Foster built his case on two main arguments: the virtues of a further diversification of the U.S. nuclear deterrent and the magnitude of the problem that defense against the cruise missiles would pose for the Soviet Union. Foster argued strenuously to the effect that, the ABM treaty notwithstanding, very great advances had been made in defenses against ballistic reentry vehicles and that future Soviet SAMs designed to counter, for example, the SRAM missile could have an ABM capability as well. In support of this theme, Foster argued elsewhere that the overriding reason for the U.S. MIRV program was not so much the Soviet Galosh ABM system around Moscow, but the ABM potential inherent in the SA-5 SAM that was being deployed on a much wider scale.[25] On the other hand, defense against a vehicle penetrating at very low altitude and at a reasonably high speed was inherently a more difficult proposition.

On this basis, Foster claimed that the SLCM could enhance the U.S. strategic deterrent in a number of ways. First of all, the threat of the simultaneous arrival of both strategic cruise missiles and ballistic missiles would increase U.S. confidence in the ability of both types of weapons to penetrate, because the Soviet Union would have to dissipate its effort on strategic defense.[26] A second dimension to his argument was that by putting cruise missiles on both bombers and submarines, the United States could prevent the Soviet Union from basing its air defense strategy on attacking U.S. bombers on the ground or intercepting them at long ranges from their intended targets.[27] Broadly speaking, a defender has to allocate the resources available for defense against bomber attack between interceptor aircraft that are useful mainly against the bomber itself, and SAMs deployed around likely targets to destroy any bomber-launched missiles. Thus if long-range cruise missiles were deployed only on bombers, the defender could efficiently concentrate its resources in long-range interceptor aircraft; but if individual missiles were also launched from submarines, the defender would have to dissipate his effort over interceptors and SAMs. In this connection Foster also alluded to the possibility of launching SLCMs as defense suppression weapons to aid bomber penetration. Third, he suggested that a seaborne strategic force composed of long-range Trident SLBMs and shorter-range Poseidon SLBMs plus cruise missiles would complicate Soviet ASW efforts by making all ocean areas relevant to such an effort. Finally, he suggested that the SLCM could "compensate somewhat" for the greater number of SLBMs permitted the Soviet Union in the SALT I agreement.[28]

With all these advantages, Foster believed that strategic cruise missiles deployed on submarines "would add more deterrent per dollar than any other of our schemes."[29] He put SLCM ahead of a standoff bomber because the latter had "a basing vulnerability that the submarine does not have at this moment."[30] He also believed SLCM had a special advantage over a mobile ICBM in penetrating the Soviet Union because of its unique flight profile and its smaller launch signature.[31] Indeed, he went so far as to suggest that deployment of the

SLCM could permit some reduction in the number of existing strategic forces.[32]

Foster neatly summarized his case for the SLCM—and confirmed that cruise missiles were discussed during SALT I—as follows:

> One, . . . it puts extra stress on the Soviet air defense environment. I think [that] is terribly important. Second, we have not been successful [in SALT] in getting the Soviets to pay any attention to their cruise missiles, which we see as a threat to our coastal populations. One way of getting their attention on this matter—and at the same time, in a credible way, redressing the imbalance that currently could result from the interim agreement—is to have a viable cruise missile program of our own. So it has both military and diplomatic connotations.[33]

INITIAL REACTIONS IN CONGRESS

Members of Congress, and others, remained skeptical. The Senate Foreign Relations Committee invited a group of defense analysts and arms controllers to testify on the SALT I agreements. None of those who volunteered an opinion specifically on the SLCM proposal considered it a good idea and one, Richard Garwin, said that it made no sense.[34] It was Senator Proxmire, however, who assembled a particularly cogent case against the SLCM, although his argument was directed as much against the B-1. Proxmire conceded that a strategic cruise missile was a useful "hedge against threats to our current strategic forces," but he argued that the SLCM was the wrong version of such a weapon.[35] The right version was a long-range, bomber-launched cruise missile. He accused the Air Force of doggedly pursuing a decoy cruise missile rather than an accurate standoff weapon in order to protect the B-1. Proxmire went on to argue that it was bombers and not SLBMs that faced penetration hazards, and that an aircraft platform for the cruise missile was far cheaper than a submarine.[36] In addition, he pointed out that the SLCM would contravene the spirit of the SALT I interim agreement because sea-based strategic weapons were limited under that agreement.[37]

The net result of this first debate was that Congress appropriated $6 million of the $20 million requested for the SLCM in FY1973. But, for SLCM supporters, things were to get far worse before they got better.

The FY1974 defense budget was presented by a new secretary of defense, Eliot Richardson. Generally speaking, Richardson reiterated the views of his predecessor, referring to the extensive cruise missile program in the Soviet Union and to the fact that these weapons were not covered by SALT I. His endorsement of the SLCM program was somewhat less emphatic than Laird's. In Richardson's words, "the United States should give some attention to this particular area of technology, for both strategic and tactical roles."[38]

DDR&E Foster, on the other hand, remained a strong supporter of the SLCM on military grounds, while other officials directly or implicitly acknowledged the role of securing bargaining leverage in the ongoing SALT negotiations. Foster's main concern continued to be that all existing strategic forces in the U.S. arsenal involved putting weapons high up in the atmosphere, thus rendering them vulnerable, in his view, to sophisticated air defenses. Putting cruise missiles on submarines would hedge against the emergence of a threat to the prelaunch survivability of strategic bombers, the existing aerodynamic leg of the Triad, and present a unique challenge to Soviet air defenses.

In responding to a question, Foster distinguished between programs intended as bargaining chips and those intended to insure the future viability of the strategic deterrent. He made it clear that, in his view, SLCM belonged in the latter category:

> A SALT chip . . . implies something that we would be willing to give up or substantially halt in exchange for a favorable concession from the other side. The B-1, SCAD, Trident and SLCM are not bargaining chips since . . . they are intended to assure the survival and penetration of our strategic forces whether the SALT agreements are continued, enlarged or reduced.[39]

To a skeptical senator or congressman it must have seemed that the Defense Department wanted to have its cake and eat it as well. When it was suggested that the SLCM proposal implied a lack of confidence in the existing Triad, the counterargument was that it was no more than a hedge against the emergence of a severe threat to one or more legs of the Triad, and as such it would not be produced and deployed unless such a threat materialized. On the other hand, when it was suggested that the SLCM was a bargaining chip, the weapon was ranked with the B-1 and the Trident as essential irrespective of the outcome of SALT.

As mentioned, other officials stressed the link with the SALT negotiations. Admiral Zumwalt, when asked why the Navy had agreed to allow its tactical cruise missile program to be relegated to a secondary position relative to the strategic weapon, responded as follows:

> The signing of the SALT agreements and the signal given to the President by Mr. Brezhnev in which he said that we

> will do everything within our limits that are not prohibited by the agreements, left us in a situation in which the Soviet Union had a very large number of cruise missiles and the United States had zero. This was a very unhealthy situation upon which they could capitalize and made it mandatory for the U.S., if we are going to be able to aspire to have a respectable SALT II agreement, to have something with which to negotiate. The intelligent way in which to do that was to combine the tactical cruise missile and the strategic cruise missile technology....[40]

The acting assistant secretary of the Navy for research and development, Dr. Waterman, was even more explicit:

> We have been asked by the Secretary of Defense, to provide a demonstration of capabilities of all the elements of the strategic cruise missile in time relationship to the SALT negotiations....[41]

The spirit of this last remark was reflected in the statements of several other Navy spokesmen during the early years of the SLCM program. That is to say, while the Navy was responsible for the development of the SLCM and defended the weapon publicly, its spokesmen discreetly but persistently reminded Congress that it was a Defense Department initiative and not the Navy's.

Another issue that arose for the first time in 1973 was contradictory assessments given by the Defense Department and the Air Force on the ability of cruise missiles to penetrate Soviet air defenses. The Department of Defense was extremely optimistic on this question; at one point Foster had stated that the SLCM's penetration capability was certainly better than bombers and at least as good as the SCAD.[42] The Air Force, on the other hand, in the face of a determined campaign by elements in Congress and the Defense Department to give the armed version of the SCAD at least as much priority as the decoy version, argued that a subsonic cruise missile would probably be highly vulnerable to terminal SAM defenses.[43] The SCAD was canceled in June 1973 in favor of the ALCM, but the Air Force stuck to its view on vulnerability right up to the cancellation of the B-1. In any event, this indication that the military community was itself divided on the capability of the cruise missile added to the already considerable pressure to drop the SLCM.

Thus while the House of Representatives agreed to authorize the full $15.2 million requested for the SLCM in FY1974, the Senate voted to deny the entire request on the ground that the Defense Department had not established a requirement for the weapon. The House-Senate

Conference settled on a compromise figure of $2.5 million, but stipulated that: neither ALCM nor SLCM could proceed beyond the development of subsystems; and the Defense Department must conduct a thorough review of cruise missiles and submit the results to Congress along with the FY1975 defense budget.

Two months later, in December 1973, the House-Senate Conference agreed to appropriate the $2.5 million for the SLCM. In the meantime, however, the House Appropriations Committee had effectively told the Navy to drop the strategic cruise missile. In its report the committee stated:

> It is doubtful that the proposed long-range strategic cruise missile is of any real military value. Ballistic missiles are inherently superior in accuracy and invulnerability. The Navy should devote its efforts to developing tactical cruise missiles, for which it has a real military requirement instead of a highly questionable strategic cruise missile.[44]

SUBSEQUENT DEVELOPMENTS

Having managed to keep the SLCM proposal alive for two years, the Defense Department now began to have more success, at least in terms of securing congressional support to proceed with advanced development. The review of the various cruise missile proposals, called for by Congress, reconfirmed the utility of these weapons, although there appears to have been some hesitation on the part of the Defense Department regarding the strategic version of the SLCM. The Navy budget for FY1975, submitted to the Defense Department in the latter part of calendar 1973, included $60 million for the SLCM. According to the SLCM project manager, Captain Locke, the Navy was asked to revise its request on the assumption that only the tactical version would be developed. The resulting figure, $45 million, was eventually included in the budget submitted to Congress; but in the meantime the deputy secretary of defense had decided to proceed with both versions.[45] Apart from SALT considerations, someone presumably reminded the deputy secretary that the Navy had already openly stated that its requirement for the tactical SLCM was not sufficiently strong to justify the development costs on its own. In other words, if the strategic variant was dropped, there was a substantial risk that Congress would feel justified in canceling the entire program.

In February 1974 the first DSARC meeting on SLCM approved the issue of contracts for the development of prototypes. The Navy selected General Dynamics and LTV as competitors during this stage of the development sequence. The winner, to be selected early in 1976, would proceed with advanced engineering development. The competitive prototypes were designated the ZBGM-109 (General Dynamics) and the ZBGM-110 (LTV).[46] The contractors were apparently told that the objective for range was a minimum of 1,400 n.m., but the Navy was already confident that this could be achieved even if the weapon flew the entire distance at low altitude.[47]

Prior to the submission of the FY1975 defense budget to Congress, James Schlesinger replaced Richardson as secretary of defense and Malcolm Currie took over from John Foster as DDR&E. Schlesinger is identified with his controversial but generally successful effort to formally commit the United States to a strategic posture that included options for limited counterforce attacks on the Soviet Union. As far as one can tell from the public record, Schlesinger was neither particularly excited by nor concerned about strategic cruise missiles. For example, when asked explicitly by Senator McIntyre what the United States was doing to preserve the "delicate balance of deterrence," they were not mentioned in the list he provided in response to the senator's question.[48] However, Schlesinger was persuaded of the utility of cruise missiles for tactical roles and pressed for them strongly on this ground.

Currie, on the other hand, was far more positive on strategic cruise missiles and, as DDR&E, he was to preside over the development of these weapons for four crucial years. During the hearings on his nomination as DDR&E, Currie expressed the view that if the performance characteristics projected for cruise missiles—range, pay load, low altitude capability, and small electronic signature—could be realized, then these weapons would be valuable in a number of roles. He further stated the following:

> I think it is unfortunate that this missile has in the past been uniquely identified with a submarine platform. I feel that we should pursue this as a technology validation program for a subsonic cruise missile, with the question of specific roles and launch platforms—land, sea or air, left open until the technology is demonstrated.[49]

In his first report to Congress early in 1974, Currie amplified on the role envisaged for SLCM. Currie characterized cruise missiles generally "as a major alternative approach to penetration of formidable Soviet air defenses."[50] From some points of view this was an unfortunate choice of words because it immediately directed attention to the alternatives that could be given up if cruise missiles worked. The focus here, of course, was the B-1.

Currie made three further salient points. First, he confined SLCM's role to the attack of "perimeter targets."[51] This may have reflected his uncertainty at that time about achieving the range necessary for deep penetration or a reluctance to require the launching submarine to move in close to the Soviet landmass. It may also have reflected a degree of skepticism on the ability of SLCM to survive sophisticated air defenses. Second, as a supplementary justification for SLCM, he pointed out that:

> it provides for the proliferation of the strategic submarine force, because every strategic and tactical submarine could become a potential launch platform for strategic cruise missiles.[52]

The Navy, for its part, went even further and listed both the possibility of covert deployment and the inability to verify whether a cruise missile was strategic or tactical as major advantages of the SLCM.[53] Objections that the program seemed to be designed to maximize the difficulties for strategic arms control were countered by the argument that the Soviet Union had already created these difficulties.

Finally, Currie argued—in the question-and-answer session rather than in his prepared statement—that cruise missiles made good strategic sense because: defense against them was an enormously expensive undertaking; being traditionally defense-minded, the Soviets would take up the challenge; and this would be to the advantage of the United States because it would divert Soviet resources away from additional offensive strategic capabilities.[54] This line of reasoning became increasingly dominant in Currie's support for cruise missiles.

As outlined by the Navy, the SLCM program for FY1975 and future years envisaged a weapon with both tactical and strategic configurations, with the primary launch platform being the nuclear hunter-killer submarines. Neither the dedicated cruise missile submarine nor SSBNs converted to launch cruise missiles vertically were being considered any longer. The program did provide for a demonstration of the feasibility of launching cruise missiles from surface ships, and it was pointed out that this would effectively validate the feasibility of a ground-launched version as well.[55] The Navy freely acknowledged that if there were no strategic program to support the bulk of the development costs it would not be pursuing a tactical variant. The latter was only attractive as an inexpensive spin-off from the strategic weapon.[56] In fact, even for the strategic weapon, an important part of the rationale was that the various advantages it offered could be secured at a relatively low cost. As Captain Locke put it: "If this were a very expensive system it would probably not be worth developing."[57]

This is additional evidence that the need for the SLCM in terms of preserving the viability of the U.S. strategic deterrent was less than compelling, and that more subjective considerations like the SALT negotiations and third-party perceptions of the strategic balance were predominant.

While Congress—this time the Senate Appropriations Committee—insisted on another in-depth review of the need for cruise missiles, a healthy $38 million was appropriated for the SLCM, only $6 million less than requested. It appears Soviet authorities also became convinced during 1974 that the SLCM was not about to go away. An article in Krasnaya Zvezda of 30 May 1974 pointed out that the development of the Harpoon missile addressed a number of technical problems associated with "strategic winged missiles." It also opined that the SLCM was "by no means just talk."[58]

This interest apparently developed quite rapidly. Currie reported that at the meeting between President Ford and General Secretary Brezhnev at Vladivostok in October, Soviet officials expressed "some deep concerns" on strategic cruise missiles to their U.S. counterparts.[59]

In November 1974 the SLCM proposal was formalized as an operational requirement (OR) by the chief of naval operations. The OR is a key document in the weapon development sequence and, although it contained nothing new, the justification advanced for the SLCM is worth quoting:

> To provide U.S. forces with a highly survivable strategic weapon system which augments existing systems by providing a diversified strategic arsenal. The SLCM would provide a hedge against the development of effective threats to pre-launch survivability of land-based forces, to inflight vulnerability of ballistic missiles and to bomber survivability. Likewise, SLCM's will complicate the Soviet air defense problem and require them to expend additional resources to counter the additional threat.[60]

In February 1975, after the complete reappraisal of cruise missile programs requested by Congress, Currie advanced a distinctly new rationale for what by then was being called the sea-launched rather than just a submarine-launched cruise missile.

> A sea-launched cruise missile development provides a desirable augmentation of capability, a unique potential for unambiguous, controlled single-weapon response and an invulnerable reserve force.[61]

The need for an invulnerable reserve force had never been officially expressed before, but Congress did not react to it in a major way and subsequent testimony provided little additional information.[62] Interestingly enough, the annual report of the secretary of defense contained almost identical wording on cruise missiles except for the reference to an invulnerable reserve force.

The probable reason for this characterization is that the Navy's description of how the SLCM would be employed left it effectively as a reserve weapon. As has been seen, around 1974 the Navy had settled on the SSN as the launch platform for the SLCM. Moreover, it was emphasized that SSNs would not, under normal circumstances, be diverted from their primary mission: antisubmarine warfare. Navy studies indicated that with new-technology ASW torpedoes, up to 6 of the 21 or 22 torpedo spaces available on an SSN could be allocated to SLCMs without degrading its ASW capability.[63] Since both the submarine-launched version of the Harpoon and the tactical version of the SLCM would be competing for these six spaces, the normal payload of strategic SLCMs would be something less. One official referred to "a couple" of strategic SLCMs per SSN.[64]

The important point, stressed in 1974 and to some extent even earlier, was that the SSNs would not be an alert strategic force. As Captain Locke put it, identifying himself very closely with the SLCM:

> We are not a first or second strike weapon. We are a deterrent, and perhaps the last deterrent.[65]

Currie went on to endorse John Foster's view that strategic cruise missiles were "an option for countering a perceived imbalance in numbers of strategic forces." He also pointed out that, as a reserve weapon, the developing debate on whether it could penetrate Soviet air defenses was largely irrelevant because of the extensive damage inflicted in preceding U.S. retaliatory strikes.[66] The reference to the cruise missile's capacity for "unambiguous, controlled single-weapon response" was clearly an attempt to find a slot for the weapon in the so-called Schlesinger doctrine. An element of contradiction arises here in that such controlled responses were envisaged as enhancing the deterrent against a full-scale nuclear exchange, so that this role would pit the cruise missile against an undamaged and presumably fully alerted Soviet air defense system.

The question of whether an SLCM force would constitute a new, fourth leg of the strategic forces proved an awkward one for defense officials. The Triad—the ICBMs, SLBMs, and bombers—was virtually a sacrosanct structure. Just as a move to a diad or a monad would be regarded as unthinkable, any suggestion that the Triad was inadequate would be highly controversial. Senator Symington, in April 1973, was one of the first to raise this question. Foster, to whom the question was directed, avoided a direct answer and expounded on his preference for greater diversity in the strategic forces.[67] These evasive tactics were employed on a number of other occasions, one of which is worth quoting. The exchange took place between Smith, a staff member of the R&D subcommittee of the Senate Armed Services Committee, and Tyler Marcy, assistant secretary of the Navy for R&D:

> Mr. Smith: Is this [SLCM] the supplement to one arm of the Triad, or is it another arm?
>
> Mr. Marcy: That is a philosophical question. . . . But it seems to me that it is clearly a supplement to one part of the Triad. And it is a new tool, so to speak. But it is not a fourth leg of the Triad.[68]

Apart from these semantics on the rationale for the SLCM, a number of other pertinent developments occurred during 1975. In addition to SSNs, the chief of naval operations designated all cruisers as platforms for the SLCM, particularly the newly proposed nuclear-powered strike cruiser. This particular proposal was not well received and for good reasons. Using a surface ship as a strategic platform would violate the fundamental requirement of minimizing the vulnerability of such platforms. The Navy's case appeared to rest on no more than the fact that the Soviet SSN-3 missile was also deployed on surface ships.[69]

In addition to surface ships, the Navy had been instructed to insure that the SLCM would be capable of being deployed on B-52 strategic bombers. A move to cancel the ALCM in favor of an air-launched version of the SLCM—usually labeled TALCM, for Tomahawk ALCM—had been overruled at two DSARC meetings held in December 1974 and March 1975. The decision was to retain both weapons, at least through the advanced development phase, mainly on the ground that constraints on size and shape dictated by the respective launch platforms were sufficiently different to justify this. However, the TALCM option was left open to the extent that the ALCM program was restructured so that it would complete advanced development at the same time as the SLCM. The House of Representatives, apparently persuaded by the Navy's claim that the SLCM was "the United States' only cruise missile . . . designed with a true air launched stand-off capability," expressed its dissatisfaction with these decisions by denying all the funds requested for the ALCM.[70] However, funding for the ALCM was subsequently restored in the House-Senate Conference. For the SLCM, $130.9 million was appropriated for the 15 months July 1975 to September 1976, only $13 million less than requested.

Finally, the ongoing SALT negotiations had begun to focus on cruise missiles. The Soviet Union was already pressing for a 600-km. range limitation on air-launched cruise missiles but, in June 1975, with the negotiations scheduled to resume on 2 July, it further proposed that all other cruise missiles be banned under a SALT II treaty. The figure of 600 km. was a range limit agreed to at Vladivostok for air-to-surface missiles, but the agreement did not specify whether this referred to ballistic or cruise missiles or both. The intention was that air-launched missiles of greater range would count as individual delivery vehicles for the purposes of SALT.

By the time Currie presented the fiscal year 1977 program for RDT&E to the Congress, he had become noticeably more cautious on cruise missiles. He argued that these weapons would have "applications in flexible response and . . . second-strike retaliation," but no reference was made to their role as an invulnerable reserve force.[71] In addition, Currie placed a relatively greater stress on the inability of the cruise missile to penetrate sophisticated surface-to-air missile defenses. This was done primarily in the context of the escalating debate on the need for a new penetrating bomber, the B-1, and Currie had, by this time, become a firm supporter of this aircraft. Nonetheless, his general support for strategic cruise missiles remained strong because most of the targets in the single integrated operational plan (SIOP) were not heavily defended by sophisticated SAMs or, indeed, by any SAMs at all.[72] The new secretary of defense, Donald H. Rumsfeld, was relatively noncommittal on cruise missiles, although he reiterated the point stressed earlier in the debate that ALCM and SLCM were playing an important role in the SALT II negotiations.[73]

The Navy, in the meantime, had come up with yet another role for the Tomahawk. Asked what would happen if the SLCM were negotiated away in SALT II, the Navy responded as follows:

> The CNO has directed that the Tercom guidance system, currently being developed for the strategic Tomahawk and the Air Force's ALCM, be utilized for a long-range tactical Tomahawk. Tercom, with its proven accuracy, will be invaluable in the projection of power ashore in tactical situations. That long-range Tomahawk would also require use of the turbofan sustainer engine.[74]

The specific intent of this response was clearly to indicate that the Navy had important uses for the Tomahawk apart from its potential as a strategic weapon. In fact, some time before this, in late 1974 or early 1975, the Navy had reviewed its requirement for the nonstrategic version of the SLCM. The Navy concluded that it was a requirement in its own right and that it would endeavor to proceed with such a weapon even without the financial support provided by the strategic program.[75] The more important point is that, in U.S. military parlance, "projection of power ashore" has a quite specific meaning and refers to Navy and Marine Corps applications of military power for a limited period to achieve limited but specific objectives. As such this role is quite distinct from strategic, tactical nuclear and antiship applications of the cruise missile. A pertinent observation here is that the automatic construction of digitized contour maps of the kind used in the TERCOM guidance system has general applications in cartography. As one senior U.S. defense official remarked, the cost of producing TERCOM maps for any place on earth will, in the near future, be a minor consideration.[76]

On the strategic version, Navy officials referred somewhat hesitantly to its utility within the Schlesinger doctrine of flexible response.[77] The following exchange, for example, makes no mention of the SLCM in the second- or third-strike (reserve force) role.

> Mr. Smith: What kind of mission does the Navy define that a strategic cruise missile on an attack submarine would prefer?
>
> Admiral Blount: That ties in with policies that are being looked at for more flexible response. . . . My understanding is that one could have a high accuracy requirement for some specific target area with low collateral damage associated with it. And in this respect it is not really matched by other strategic weapons that we have in the arsenal.[78]

In his final report to Congress as DDR&E, Currie described the advent of long-range, highly accurate cruise missiles as "perhaps the most significant weapon development of the decade," and stressed the fact that they represented a high leverage investment: relatively inexpensive for the United States but a very costly defense problem for the Soviet Union.[79] As in his previous report, however, Currie was relatively explicit on the role and utility of ALCM, but offered no comparable justification for SLCM. The new report was different mainly in terms of the relative emphasis given to different versions of the Tomahawk. First, the tactical antiship version of Tomahawk, previously regarded as a secondary spin-off from the strategic program, was discussed on equal terms, with the latter thus endorsing the Navy's view on the importance of this version. Second, relatively greater prominence was given to launching platforms other than submarines for the land-attack version.[80]

There were other indications of a significant change in official thinking on the role of the SLCM during the preparation of the FY1978 budget (that is, in the latter months of 1976). The assistant secretary of the Navy for R&D, Tyler Marcy, stressed that the Navy's primary interest in Tomahawk was the conventionally armed antiship variant.[81] While this may be construed as a degrading of the strategic land-attack version, one must remember that the Navy consistently regarded the strategic cruise missile as an OSD development it had been directed to carry out, rather than a Navy initiative. Nevertheless, Marcy's additional comment that the land-attack version of Tomahawk was valuable, "particularly in a theater role," provided some indication that the general strategic application of SLCM was being reconsidered.[82]

It was left to the Tomahawk project manager, Captain Locke, to take the mystery out of this situation. The following exchange was probably the first explicit indication that the strategic submarine-launched cruise missile was defunct:

Mr. Ichord: Let me ask more specifically. Why would we need the strategic submarine-launched cruise missile?

Captain Locke: There is no strategic submarine-launched cruise missile planned. The only purpose that we have for the submarine cruise missile is for theater war. That is a tactical mission.

Mr. Ichord: You say that you have eliminated this, Captain Locke?

Captain Locke: I haven't. It was determined by the Navy last year, starting about this time, the CNO directed an intensive study in the use of cruise missiles for the Navy. . . . The conclusion was that the land-attack cruise missile was of use for the Navy in a theater role.[83]

Given the point referred to earlier that the Navy viewed the strategic cruise missile as an external initiative, it seems reasonable to presume that it could not be canceled by the Navy alone. In other words, the studies ordered by the CNO and the conclusion that only a theater land-attack SLCM was useful must have been reviewed and endorsed by higher authorities, notably the secretary of defense, the DDR&E, and the president or his national security advisors. As has been seen, the annual report of the DDR&E did not reflect this narrowing in the role envisaged for the land-attack cruise missile. The secretary of defense, in fact, continued to refer to a submarine-launched strategic cruise missile distinct from the theater version of this weapon.[84]

Significantly, however, the secretary of defense devoted an unusual amount of space to the artificiality of the distinction between strategic and theater or tactical nuclear weapons. He pointed out that there were a number of weapons in existence that were not unambiguously strategic or tactical, and that two recent Soviet systems—the SS-X-20 IRBM and the Backfire medium bomber—were important additions to this "gray area." He went on to argue that cruise missiles were particularly difficult to classify, and that the classification used in his report would probably require change as cruise missile technologies evolved.[85]

It seems quite clear that these open-ended comments on the roles and missions for cruise missiles were designed to give the United States room for maneuver in the SALT negotiations. In fact, it can be argued that the redirection of the SLCM from a general strategic to a theater land-attack role sometime in 1976 was designed in part to bring what the Pentagon was saying more into line with the U.S. negotiating position in SALT. That negotiating position, as described by Admiral Armstrong in March 1976, was "that the cruise missile is not a part of our central strategic system in any way whatsoever."[86]

It must be stressed that the apparent elimination, at least for the time being, of the strategic SLCM did not mean that the weapon's performance characteristics were changed in any way. It was exactly the same weapon. In Locke's words:

> It is a case of semantics. Because we started in 1972 calling all cruise missiles strategic we have been led to this confusion.[87]

Nor was any change envisaged in launch platforms; the SSN was still the leading candidate. Locke continued to stress the flexibility cruise missiles would endow on SSNs since the mix of ASW torpedoes, antiship cruise missiles, and land-attack cruise missiles could be varied according to the predominant mission assigned to an SSN on a particular patrol. The Soviet Union, being unable to determine what the mix of weapons was, would be forced to assume that all SSNs carried all types of weapons and make the appropriate investment in defenses.[88]

Further amplification on the SLCM's new role was provided by Vice-Admiral Long, deputy CNO for submarine warfare:

> The primary need for the land-attack Tomahawk is in a theater role where its single warhead, high accuracy capability with resultant low collateral damage, penetrability and survivability make it ideal for use in limited nuclear attacks as a theater weapon. It represents one of the few new systems the U.S. could deploy if needed to maintain theater balance in the face of growing Soviet peripheral attack capabilities that include such systems as the Backfire bomber and the SS-20 mobile ground launched missile. . . .
>
> Sea-based theater weapons are not tied to a particular land area, but can be deployed in relatively short time to provide a theater capability throughout the world. . . .
>
> If carried on board submarines, they constitute a uniquely survivable weapon system capable of operating undetected in areas hostile to other forces, and are suitable as a strategic reserve force if future needs dictate. . . .
>
> The increased flexibility in tasking afforded by the deployment of land-attack Tomahawks would significantly increase the options available to the National Command Authorities as a stabilizing factor in the event of conflict while leaving our strategic forces intact.[89]

In January 1977 the DSARC group met again to consider whether the SLCM and ALCM should proceed into full-scale engineering development. It was a notably productive meeting. First of all, it was decided to create a joint Navy-Air Force project office to manage the entire cruise missile development effort, with the Navy as the lead service. Engineering development was approved for all versions of the Tomahawk, that is, theater land-attack and tactical antiship (both submarine launched) and the ground-launched cruise missile that had been proposed as a separate program in the FY1978 budget.[90] The ALCM was also approved for engineering development, but the Air Force was overruled and instructed to give first priority to the extended range ALCM-B, which could not be accommodated in the B-1.

As this book is concerned with strategic cruise missiles, it is not relevant to pursue the programmatic history of the SLCM any further. The weapon's primary significance for this study is that it pioneered the renewal of U.S. interest in strategic cruise missiles and, as shall be seen in the next chapter, it survived in this role long enough to impart an unstoppable momentum to the strategic air-launched cruise missile. The nuclear-armed, land-attack SLCM remains an active program with a development schedule aiming at an initial operational capability in 1983. It should be pointed out, however, that in addition to being reclassified as a theater weapons system, the SLCM also came to be overshadowed in this role by its ground-launched variant, the GLCM.

The Air Force was given responsibility for developing the GLCM because deep-penetration strikes at the theater level is an Air Force mission. This allocation of missions stems from the Thor-Jupiter IRBM controversy in the mid-1950s over whether the Army or the Air Force should have jurisdiction over weapons that strike deep behind the enemy's front lines. The demarcation line on weapon range is not precisely fixed, but the longest-range battlefield weapon the Army currently operates is the 400 n.m. Pershing ballistic missile. It is also worth recalling that Matador and Mace were Air Force weapons.

The Ford administration's FY1978 budget requested just $3.9 million to initiate GLCM development, but in amendments to this budget, the Carter administration boosted the sum to $27.9 million.[91] The official rationale for the GLCM was at first essentially twofold: to provide a partial substitute for NATO's dual-capable aircraft assigned to deliver nuclear weapons and thus permit more aircraft to be released for conventional operations; and to complement the Army's Pershing missile and compound the Warsaw Pact's defense problems by presenting a tactical nuclear threat composed of ballistic and cruise missiles as well as aircraft.[92]

One feature of the GLCM considered particularly valuable was its high mobility, which would give it better prelaunch survivability than Pershing.[93] The Pershing is mobile but each new launch site has to be carefully surveyed to preserve the weapon's accuracy. The GLCM's TERCOM guidance allowed more flexibility in this regard. Another virtue claimed for the weapon was that it would be less costly than aircraft to maintain on Quick Reaction Alert.[94] It was also suggested that GLCMs with conventional warheads could cost-effectively replace aircraft delivering conventional munitions against heavily defended targets where aircraft attrition rates would be high.[95] This role would significantly drive up the number of GLCMs required. That the GLCM—or any cruise missile variant for that matter—could eventually achieve the very high accuracy required to permit a conventional warhead was demonstrated early in 1977 when the weapon was tested with a terminal guidance system called SMAC. A photograph reproduced in congressional hearings showed the missile missing the aimpoint by less than 15 feet. Being optical, SMAC terminal guidance would be practical only in daylight and fair weather, but alternatives—such as imaging infrared—are available to lessen this limitation.

SUBMARINE-LAUNCHED CRUISE MISSILE / 51

The defense budget for FY1979, the first prepared by the Carter administration, retained all the cruise missile programs, but the spotlight was on the air-launched version. The cancellation of the B-1 in June 1977 had made the air-launched cruise missile "our highest national priority."[96] This rather dramatic elevation in status was based on an additional line of reasoning that ran as follows. The SLBM force was by far the most invulnerable of U.S. strategic forces and, with the Trident program, the United States was moving to a relatively greater reliance on this leg of the Triad. This increasing reliance, though logical under prevailing conditions, made possible Soviet technological breakthroughs in ASW or in ABM capabilities of greater consequence, and put a high premium on hedges against the eventuality of such technological breakthroughs. Secretary of Defense Brown regarded air-breathers, bombers, or cruise missiles as the "hedge of first choice, with (especially mobile) ICBMs as an important second, against possible threats to our essential SLBM force."[97] Because the United States had to be certain of the success of this program, the Tomahawk was made a fully fledged alternative to the Air Force's ALCM, with a competitive flyoff to take place in 1979.

A major objective of the FY1979 budget was to address what Defense Secretary Brown referred to as "an increasingly precarious conventional balance between NATO and the Warsaw Pact in Europe."[98] In this context the sea-launched version of the Tomahawk was funded at a healthy level and its theater nuclear land-attack role was made explicit both by Brown and Currie's successor, William J. Perry.[99] Similarly, the GLCM version of the Tomahawk was maintained as a separate program to further augment the theater nuclear forces. Perry explicitly indicated that the GLCM was a direct and deliberate counter to the Soviet SS-20 MIRVed IRBM.[100]

During the course of the hearings on the FY1979 defense budget, the GLCM proposal began to run into some opposition, particularly in the House of Representatives, as a consequence of a more general concern that modernization of the tactical nuclear weapons stockpile was resulting in too many overlapping programs.[101] In addition to the SLCM and the GLCM, the Army, since 1973, had been developing the Pershing II. The Pershing II will have a radar area correlator terminal homing system that is expected to give the weapon a CEP of about 120 feet.[102] This is at least as good as (probably better than) the accuracy the GLCM could achieve if it relied solely on TERCOM guidance. Moreover, being a ballistic missile, the Pershing's probability of penetrating to the target is essentially unity while that for the GLCM, though very high, will be inherently lower. The GLCM's other edge in performance, its extremely long range, was at the time subject to the outcome of the negotiations on both SALT II and SALT III.[103]

This long range contributed substantially to the view that the GLCM would have greater prelaunch survivability than Pershing. The Pentagon, with the support of European NATO countries, continued to support both GLCM and Pershing. The case for the GLCM rested on cost and the familiar argument that a mixed force of cruise and ballistic missiles would compound the enemy's defense problems. In terms of cost the GLCM was presented as a very attractive proposition. Having a lower unit cost than Pershing and lower requirements in terms of manpower and support equipment, the GLCM was expected to be about three times more cost-effective in the nuclear delivery role and ten times more cost-effective in the conventional role.[104]

Presumably in consequence of these results, a specialized conventional variant of the GLCM made its debut in mid-1978. This weapon—called TAAM or CAAM for Tomahawk or conventional airfield attack missile—would employ TERCOM guidance plus SMAC terminal homing, and carry a warhead loaded with submunitions capable of penetrating and cratering runways. In this role some 78 inches of the missile's airframe—with 13 cubic feet of space—would be devoted to the payload, which suggests that the weapon's range has been quite severely cut back.[105] In its nuclear version the GLCM still had the political factor in its favor, that is, a long-range weapon to be deployed as a direct countermove to the Soviet SS-20.

This concern with political or perceptual considerations may help to explain why the Air Force chose this time to dust off its mobile medium-range ballistic missile, a project canceled in 1964. This weapon would have a range capability of 500 to 1,200 n.m. The FY1979 defense authorization bill provided $2 million for initial design work.[106] With the Congress already split over the existing tactical nuclear weapon program, considerations other than purely technical ones must have motivated the decision to resurrect this concept. A pertinent observation here is that the Army had deliberately retained the 400 n.m. range for the Pershing II to avoid the political implications of a longer-range system that could strike targets on Soviet territory.[107] This was done despite the fact that more advanced rocket engines and propellants were available that could give the Pershing II a range of approximately 1,000 miles. With the evident demise of the political inhibitions concerning long-range theater missile systems, the Army chose to exercise this option and the Pershing IIXR (for extended range) became a third contender for the requirement to respond to the SS-20.[108]

The defense budget for FY1980 requested funds to complete development of the nuclear-armed, land-attack SLCM, but no monies were requested for procurement.

For the GLCM, two development schedules were proposed, one permitting initial deployment in FY1984 and the other moving this forward to FY1982.[109] In January 1979 it was casually reported that the Air Force had lost interest in conventionally armed cruise missiles and that the Navy was now pursuing such a variant of its land-attack SLCM.[110]

In July 1979 these various developments distilled into a formal U.S. proposal to its European NATO allies to deploy a mixed force of GLCMs and Pershing IIXRs as a response to the Soviet SS-20 IRBM. After a long and controversial process of review, the NATO countries decided in December 1979 to proceed with this program. The intention now is to begin deployment in 1983 of 108 Pershing IIXRs (all to be deployed in the Federal Republic of Germany) and 464 GLCMs (160 in the United Kingdom, 112 in Italy, 96 in Germany, and 48 each in Belgium and the Netherlands).[111]

NOTES

1. Ron Huisken, "World Stock of Fighting Vessels," in World Armaments and Disarmament, SIPRI Yearbook 1975 (Stockholm: Almqvist and Wiksell, 1975), p. 305.

2. Ibid., pp. 304-5.

3. A computation of the capital value of the two fleets—allowing for the size, vintage, armament, and type of propulsion of individual ships—put the Soviet fleet at 41 percent of the U.S. fleet in 1960 (ibid., p. 297).

4. Fiscal Year 1976 and July-September Transitional Period Authorization for Military Procurement, hearings, Senate Armed Services Committee, Part 10, April 1975, p. 586.

5. Fiscal Year 1972 Authorization for Military Procurement, hearings, Senate Armed Services Committee, March-May 1971, p. 945.

6. The Harpoon missile itself is more slender than a torpedo-tube, so it must be encased in a torpedo-size capsule to permit launching from submarines. This version is sometimes called Capoon for obvious reasons.

7. Fiscal Year 1973 Authorization for Military Procurement, Addendum No. 1: Amended Military Authorization Request Related to Strategic Arms Limitation Agreement, hearings, Senate Armed Services Committee, June-July 1972, p. 4,353.

8. This will be discussed in more detail below.

9. Fiscal Year 1972 Authorization for Military Procurement, hearings, p. 2,732.

10. Fiscal Year 1977 Authorization for Military Procurement, hearings, Senate Armed Services Committee, Part 6, April 1976, p. 3,365.

11. Ibid.
12. Hearings on Military Posture and HR 12604, House Armed Services Committee, April 1972, p. 10,744.
13. Fiscal Year 1973 Authorization for Military Procurement, hearings, Senate Armed Services Committee, March-May 1972, pp. 583ff.
14. Fiscal Year 1974 Authorization for Military Procurement, hearings, Senate Armed Services Committee, Part 4, March 1973, p. 2,635.
15. Fiscal Year 1973 Authorization for Military Procurement, Addendum No. 1, hearings, p. 4,244.
16. The Operational Requirement is perhaps the most important document in the formal genesis of a weapons system in the United States. The document for SLCM is reprinted in Fiscal Year 1977 Authorization for Military Procurement, hearings, pp. 3,369-72.
17. Fiscal Year 1975 Authorization for Military Procurement, hearings, Senate Armed Services Committee, Part 7, April 1974, p. 3,652.
18. Fiscal Year 1978 Authorization for Military Procurement, hearings, p. 2,680.
19. Fiscal Year 1973 Authorization for Military Procurement, hearings, p. 583.
20. These were the site defense ABM ($60 million), bomber rebasing ($45 million), verification capabilities ($13 million), improved reentry vehicles for ICBMs and SLBMs ($20 million), and Command Control and Communications ($10 million).
21. Fiscal Year 1974 Authorization for Military Procurement, hearings, p. 159.
22. These views are examined in more detail in Chapter 5.
23. Fiscal Year 1973 Authorization for Military Procurement, hearings, p. 4,244.
24. "Kissinger Critique," Economist, 3 February 1979, pp. 17-18.
25. Hearings on Cost Escalation in Defense Procurement Contracts and Military Posture and HR6722, House Armed Services Committee, April 1973, p. 2,844.
26. Fiscal Year 1973 Authorization for Military Procurement, Addendum No. 1, hearings, p. 4,340.
27. Ibid., p. 4,297. As we shall see below, Foster consistently advocated that the Air Force reorient its program for a decoy missile (the SCAD) to an armed standoff weapon.
28. Ibid., p. 4,340.
29. Ibid., p. 4,356.
30. Ibid.
31. Ibid., p. 4,369. The launch of an ICBM generates a very

considerable amount of heat and light energy that can be detected by space-based early-warning systems. The SLCM, on the other hand, would use a relatively tiny rocket booster to propel it out of the water, and its turbofan sustainer engine generates very little heat. Just how revealing the launch of an SLCM would be is an aspect on which very little has been said. Since the booster is ignited under water and fires until the missile is at an altitude of 1,000 to 1,500 feet, one can presume that considerable acoustic energy is generated (detectable by sonar over great distances) and that the glare is visible from some distance.

32. Ibid., p. 4,367.
33. Ibid., p. 4,356.
34. Strategic Arms Limitation Agreements, hearings, Senate Foreign Relations Committee, June–July 1972, p. 356.
35. Quoted in Space Business Daily, 6 July 1972, p. 18.
36. It will be recalled that a new class of cruise missile submarines was still an option at this stage, although not preferred by either the Navy or the Defense Department. The option was dropped early in 1974 after preliminary design of the submarine had been completed.
37. Space Business Daily, 6 July 1972, p. 19.
38. Hearings on Military Posture and HR 6722, p. 334.
39. Fiscal Year 1974 Authorization for Military Procurement, hearings, p. 1,000.
40. Ibid., p. 1,648.
41. Ibid., p. 3,274.
42. Fiscal Year 1973 Authorization for Military Procurement, Addendum No. 1, hearings, p. 4,353.
43. Fiscal Year 1974 Authorization for Military Procurement, hearings, p. 3,537.
44. Quoted in A. A. Tinajero, Cruise Missile (Subsonic): U.S. Programs, Congressional Research Service, 1B76018, 16 September 1977, p. 19.
45. Fiscal Year 1975 Authorization for Military Procurement, hearings, p. 2,472.
46. The letter "Z" is used to identify a vehicle in the planning and predevelopment phase; "B" means the weapon can be launched from a variety of platforms; "G" indicates the weapon will be used to attack surface targets on land or at sea; and "M" indicates it is a missile.
47. Fiscal Year 1975 Authorization for Military Procurement, hearings, p. 3,658.
48. Ibid., p. 280.
49. Nominations of Mendolia, McClay, Currie and Bowers, hearings, Senate Armed Services Committee, 14 June 1973, p. 14.

56 / ORIGIN OF THE STRATEGIC CRUISE MISSILE

50. Department of Defense Appropriations for 1975, hearings, House Committee on Appropriations, April 1974, p. 401.
51. Ibid., p. 447.
52. Ibid., p. 462.
53. Fiscal Year 1975 Authorization for Military Procurement, hearings, p. 3,620.
54. Department of Defense Appropriations for 1975, hearings, p. 580.
55. In view of this the Army designated a liaison officer to the SLCM project office, but as it turned out it was the Air Force that was given responsibility for the ground-launched version.
56. Fiscal Year 1975 Authorization for Military Procurement, hearings, pp. 3,677-78.
57. Ibid., p. 3,639.
58. Quoted in Space Business Daily, 17 June 1974, pp. 256-57.
59. Fiscal Year 1976 and July-September Transitional Period Authorization for Military Procurement, hearings, p. 5,180.
60. Reproduced in Fiscal Year 1977 Authorization for Military Procurement, hearings, p. 3,370.
61. The Department of Defense Program of Research, Development, Test and Evaluation, FY1976, p. 78.
62. See James R. Schlesinger, secretary of defense, Annual Defense Department Report, FY1976 and FY1977, 5 February 1975, pp. 11-39.
63. The torpedo capacity of an SSN is based on a comment by a Navy official that one torpedo represents "5 percent or something less" of the total (Fiscal Year 1977 Authorization for Military Procurement, hearings, p. 3,366). The figure six is given in Fiscal Year 1978 Authorization for Military Procurement, hearings, Senate Armed Services Committee, February-March 1977, p. 607.
64. Fiscal Year 1975 Authorization for Military Procurement, hearings, p. 3,658.
65. Ibid., p. 3,680.
66. Fiscal Year 1976 and July-September Transitional Period Authorization for Military Procurement, hearings, pp. 2,893, 3,052.
67. Fiscal Year 1974 Authorization for Military Procurement, hearings, p. 973.
68. Fiscal Year 1977 Authorization for Military Procurement, hearings, p. 3,367.
69. The cruiser/SLCM proposal also provides an example of the importance assigned to having visible military power, even if military logic must be compromised. To quote Captain Locke: "The Soviets put their cruise missiles right out where everybody can see them. . . . We considered burying them in the superstructure where they would be a little less vulnerable to shrapnel or a hit. Instead,

we are putting them out where they will be exposed, but visible. The purpose is to make them visible and match the philosophy that the Soviet Union has followed" (Fiscal Year 1976 and July-September Transitional Period Authorization for Military Procurement, hearings, pp. 5,185-86).

70. Ibid., p. 5,130.

71. The Department of Defense Program of Research, Development, Test and Evaluation, FY1977, p. III-11.

72. Ibid., pp. III-26-27. The SIOP is the strategic nuclear war plan. It identifies the targets to be attacked with strategic weapons, allocates particular weapons to particular targets, species the time sequence of attack, and so on.

73. Donald H. Rumsfeld, secretary of defense, Annual Defense Department Report FY1977, 19 January 1976, p. 69.

74. Fiscal Year 1977 Authorization for Military Procurement, hearings, p. 3,333.

75. Fiscal Year 1976 and July-September Transitional Period Authorization for Military Procurement, hearings, p. 5,179.

76. John Walsh, deputy director, strategic and space systems, office of the director of defense research and engineering. See his testimony in Fiscal Year 1978 Authorization for Military Procurement, hearings, p. 6,435.

77. This is also true of John Walsh, Currie's deputy director for strategic and space systems. See Hearings on Military Posture and HR22500, p. 188.

78. Fiscal Year 1977 Authorization for Military Procurement, hearings, p. 3,366. Smith was a staff member of the Senate Armed Services Committee and Admiral Blount was director of the undersea and strategic warfare development division.

79. "The Department of Defense Program of Research, Development, Test and Evaluation, FY1978," reprinted in Hearings on Military Posture and HR5068, pp. 1,631-32.

80. Ibid., p. 1,633.

81. Fiscal Year 1978 Authorization for Military Procurement, hearings, p. 5,636.

82. Ibid.

83. Hearings on Military Posture and HR5068, p. 1,099.

84. Donald H. Rumsfeld, secretary of defense, Annual Defense Department Report, FY1978, p. 56.

85. Ibid., pp. 55-57.

86. Fiscal Year 1977 Authorization for Military Procurement, hearings, p. 3,367.

87. Hearings on Military Posture and HR5068, p. 1,099.

88. Ibid., p. 1,100.

89. Department of Defense Appropriations for 1978, hearings,

House Committee on Appropriations, Part 2, February-March 1977, p. 235.

90. For a short time the GLCM was labeled LLCM (land-launched cruise missile), but the literal expression of this acronym, "lickcum," was considered inappropriate.

91. The additional monies were taken from the allocation to the antishipping variant of the Tomahawk on the ground that the Navy's ability to detect and classify over-the-horizon targets was lagging behind the 300 n.m. range of this weapon. For a list of the various programs under way to rectify this deficiency, see my "The Tomahawk Anti-Shipping Missile," Pacific Defence Reporter (July 1978): 39.

92. See testimony by Secretary of Defense Harold Brown, Department of Defense Appropriations for 1978, hearings, pp. 104-5.

93. Ibid., p. 333.

94. Fiscal Year 1978 Authorization for Military Procurement, hearings, p. 6,429.

95. Ibid., pp. 6,427-28.

96. Harold Brown, secretary of defense, Department of Defense Annual Report, Fiscal Year 1979, p. 119.

97. Ibid., p. 59.

98. Department of Defense Annual Report, Fiscal Year 1979, p. 3.

99. Ibid., p. 133, and William J. Perry, undersecretary of defense, research and engineering, The FY1979 Department of Defense Program for Research, Development and Acquisition, pp. V-19-21.

100. The FY1979 Department of Defense Program for Research, Development and Acquisition, p. V-21.

101. It can be mentioned here that, despite congressional resistance, the Pentagon continued to express interest in a nuclear armed land-attack SLCM launched from surface ships.

102. Aviation Week and Space Technology, 15 May 1978, p. 20. It was further stated that this high accuracy would permit a warhead "substantially smaller" than the 60-400 KT variable yield device currently carried by the Pershing.

103. The protocol to SALT II, valid until December 1981, limits deployed GLCMs to a range of 600 km., though testing is permitted to substantially longer ranges. It was presumably hoped that some bargain could be struck in SALT III before the protocol expired or before long-range GLCMs were deployed in large numbers. Being almost identical to the SLCM, the GLCM has a range capability of about 3,800 km.

104. Aviation Week and Space Technology, 19 June 1978, p. 29.

105. Ibid. In the nuclear version of the Tomahawk, the warhead

takes up 41 inches of the airframe. The Pershing II is also a contender for the CAAM mission.

106. Aviation Week and Space Technology, 15 May 1978, p. 18, and International Herald Tribune, 4 August 1978, p. 1. The budget for FY1980 requested $4 million to continue the work (Aviation Week and Space Technology, 29 January 1979, p. 18).

107. Fiscal Year 1978 Authorization for Military Procurement, hearings, p. 6,420.

108. Flight International, 15 May 1978, p. 1,846.

109. Aviation Week and Space Technology, 9 October 1978, p. 18.

110. Aviation Week and Space Technology, 8 January 1979, p. 11.

111. For more details on this program, see "Eurostrategic Weapons," World Armaments and Disarmament, SIPRI Yearbook 1980 (London: Francis and Taylor, 1980), pp. 175-86.

4

THE AIR-LAUNCHED CRUISE MISSILE: PENETRATING VERSUS STANDOFF STRATEGIC BOMBERS

INTRODUCTION

From the late 1940s and throughout the 1950s, long-range bombers constituted virtually the whole of the U.S. strategic deterrent. The Navy provided some support with a capacity to attack peripheral targets on the Soviet landmass with carrier-borne aircraft and, from 1955, with the submarine-launched Regulus I cruise missile. At its peak around 1960, the Strategic Air Command possessed nearly 2,000 bombers, a mixture of B-47 and B-58 medium-range aircraft and the long-range B-52.[1] From 1958, with the deployment to Europe of Thor and Jupiter IRBMs, and particularly from 1959-60 with the advent of U.S. ICBMs and SLBMs, bombers came to be viewed as the air-breathing component of the strategic Triad.

Although the relative importance of bombers in the U.S. strategic force has declined markedly, they remain very important in absolute terms. In January 1978 the strategic bomber force consisted of 349 B-52s and 66 FB-111s. This force accounts for a disproportionately large share of the budget for strategic forces and disposes of "approximately one-third of all deliverable U.S. strategic nuclear weapons and half the total megatonnage."[2]

If one counts the programs terminated during development, the strategic bomber-launched air-to-surface missile (ASM) is by no means a rare species in the United States. All these weapons have had either or both of the following roles: to enable the bomber to attack additional targets, particularly those some distance from the approach route to the aircraft's primary target or targets; and to suppress defenses and enhance the bomber's ability to penetrate to its primary target or targets. The latter has probably always been the predominant motive. By the mid-1950s the vulnerability of the high altitude bomber was already an issue. In 1953 and 1954 long-range surface-to-air missiles had become operational in the Soviet

Union and the United States respectively. Although the Americans knew very little about the Soviet weapon (the SA-1 Guild), they could use the capabilities of their own (the Nike Ajax) as a guide.[3] As already noted, the decision to use low-level penetration as an optional mode of attack for strategic bombers was taken as early as 1957, the year in which the SA-2 Guideline, a second-generation Soviet SAM, became operational.

One of the earliest efforts to produce a bomber-launched ASM was a weapon known as the GAM-67 Crossbow. The Crossbow was a subsonic, turbojet-powered weapon weighing about 1,140 kg. at launch. It was designed to home in on enemy radars at ranges up to 320 km. and was planned for initial operational capability (IOC) in February 1956. In 1954 the Air Force added the requirement for a nuclear warhead, but the weapon was canceled in November 1956. Development of the radar seeker was continued and, in June 1957, a general operational requirement (GOR) was issued for a successor weapon known as Longbow.[4] The Longbow was also terminated after a relatively brief development effort.

Another weapon was the GAM-63 Rascal. The origins of Rascal go back to a study program for a bomber penetration missile in April 1946. Work actually started in 1951 with the objective of developing a rocket-powered missile with a range of 175 km., a 1,364-kg. nuclear warhead, and a circular error probable at 120 km. of 1,500 ft. The launching aircraft would locate the target and launch the missile in the right direction. In the terminal stages of its flight, an active radar seeker in the missile would provide guidance.[5] In 1957 Rascal, nicknamed "crew saver," became the first standoff bomb to reach the operational trials stage, but the program was canceled in 1958.[6] Apart from acute reliability problems, the demise of the Rascal was probably due in part to the ongoing Hound Dog program. The Hound Dog promised to be at least as fast as Rascal (Mach 1.5 or better), had five times the range, and was more than 20 percent lighter at launch.

A third program was a weapon known as the Wagtail. Development started in September 1956 and reportedly successful tests were conducted in 1958, but the weapon was canceled in 1960. The following quotation essentially sums up the publicly available information on the Wagtail:

> All that is known officially is that the Wagtail had a solid-propellant rocket motor. Unofficial reports have stated that forward-firing rockets slow it after launching, before its sustainer fires, so that its inertial guidance system and sensing devices can steer it over or around obstacles such as hills when it is launched from a bomber at very low altitude.[7]

The most ambitious attempt to increase the versatility and prolong the viability of the strategic bomber in this early period was the SGAM-87A Skybolt (later redesignated the XGAM-48A). Skybolt was a two-stage, hypersonic ballistic missile with a range of 1,850 km. Development was started in May 1959 and IOC was scheduled for 1964 but, in a controversial move, President Kennedy and Secretary of Defense McNamara canceled the weapon in December 1962. The Skybolt story is a complex one, but the proximate grounds for its cancellation were high technical risk and the U.S. move away from a counterforce/damage limitation nuclear strategy to the less demanding strategy of assured destruction.

Two other systems, both decoys, must be mentioned in this introductory survey of U.S. bomber-launched missiles. The first is the GAM-72A Quail, a decoy missile developed between 1955 and 1961 and still operational on B-52s in 1978. The second is the SN-73 Goose (also known as Bull Goose and Blue Goose). The Goose was inertially guided, had a design range of 3,200 km., and was intended to penetrate in advance of the bombers to confuse and dilute defenses. Although it was launched from the ground, it is mentioned here because of the active consideration given to putting a warhead on some of the missiles.[8] Goose was canceled in 1958, but the issue of arming a decoy missile arose again in connection with subsonic cruise armed decoy and played a significant part in the latter's demise.

Two conclusions might be drawn from this brief survey. On the one hand, once could assume that the Air Force had a sustained interest in bomber-launched missiles or, alternatively, was seriously concerned over the ability of strategic bombers to penetrate Soviet airspace. On the other hand, the high proportion of cancellations could suggest that the Air Force was not anxious to push missile developments to the point of undermining the need for long-range penetrating bombers. In any event, throughout the 1960s the Air Force had to contend with strong and effective pressure from top civilian officials in favor of bomber-launched missiles rather than continued refinement of the concept of the manned penetrating bomber.

The emergence of the ALCM as a long-range standoff missile is linked, through the SCAD decoy, with the long delay in initiating the development of a successor to the B-52. It is not a coincidence that this delay exactly matched Robert McNamara's term of office as secretary of defense. When McNamara assumed his new position early in 1961, two strategic bombers were in production—the B-52 and the B-58—and a third, the B-70, was under development. McNamara was of the opinion that the viability of the penetrating strategic bomber, most particularly the high altitude penetrator, was very much open to question. One of his early actions was to finish off the B-70. One has to say "finish off" because the B-70 was effectively canceled in

December 1959 but, in ways and for reasons that are not germane to this discussion, it reemerged in October 1960 with full weapons system status as a launching platform for Skybolt.[9] What is central to this discussion is the arguments used by McNamara to defeat supporters of the B-70, and to do so without promising to initiate the development of a substitute bomber. In January 1962 he informed Congress that studies had shown that the speed and altitude advantage that the B-70 had over the B-58 and B-52 had no significant effect on its vulnerability to SAMs, that the aircraft's performance at low altitude was no better than existing systems, and that it had not been designed to carry air-to-surface missiles.[10] Two months later McNamara expanded on the subject when the House Armed Services Committee voted monies for a reconnaissance-strike version of the B-70, the RS-70. The proposed mission for the RS-70 was to penetrate the Soviet Union after a U.S. retaliatory attack with ICBMs and SLBMs for purposes of reconnaissance, damage assessment, and to strike any remaining targets. McNamara's response was that:

> We think that the B-52's and B-58's, arriving after our missiles have suppressed the enemy's air defense, could penetrate as well, or almost as well, as the RS-70 [and that the RS-70 would] require the development of new air-launched strike missiles [that] because of their limited size and warhead yield would have to be far more accurate than any strategic air-launched missile now in production or development.[11]

McNamara never wavered from this position. In 1963 the Air Force tabled a new proposal for a strategic bomber—the advanced manned strategic aircraft (AMSA). In concept AMSA moved away from an emphasis on speed over the target to such things as quick reaction, a short takeoff run, and defensive avionics, but the Office of the Secretary of Defense consistently refused to support proposals to initiate development. The systems analysis division within OSD, under the direction of Alain Enthoven, reported on studies that indicated that bomber speed was not particularly important, while penetration aids and standoff missiles could be very important.[12] The continuing delays on the AMSA, among other things, caused a noticeable deterioration in the relationship between the secretary of defense and the congressional armed services committees. Nevertheless, McNamara steadfastly insisted that "manned bombers of the future are simply going to be launching platforms for missiles."[13]

64 / ORIGIN OF THE STRATEGIC CRUISE MISSILE

THE ARMED DECOY CRUISE MISSILE

As the earliest possible deployment data for a new strategic bomber receded into the future, the Air Force began to argue that the B-52 would become vulnerable to projected air defenses in the mid-1970s and that some steps had to be taken to enhance its ability to penetrate. Two particularly important steps were taken to this end: first, the decision was taken in fall 1968 to equip all late-model B-52s with the SRAM (short-range attack missile); and second, studies were initiated for a new decoy missile to replace the Quail. The requirement for a new decoy emerged from a series of bomber penetration studies conducted by the Air Force in 1966-67, and in all probability this is also true of the B-52-SRAM combination. The decision to develop the SRAM was taken in 1964 and development started in 1965. It was widely assumed that SRAM would be matched with the AMSA, but this was not to be. In December 1965 McNamara announced that SRAM would be deployed on a bomber version of the controversial TFX tactical fighter, an aircraft that subsequently emerged as the FB-111A.

It is likely that McNamara viewed this decision as a relatively inexpensive way of dissipating the pressure for a new strategic bomber, but the Air Force and its supporters never regarded the FB-111A as more than a stopgap or interim weapon. A small digression is worthwhile at this point: if McNamara was so convinced that the manned, penetrating strategic bomber had distinct dinosaurian qualities, why did he not do more in the area of air-launched standoff missiles? The SRAM was an important development but its range (100 nautical miles) in no way invalidated the concept of a penetrating strategic bomber. Other possibilities clearly existed. The discussion in Chapter 3 suggests that, by the mid-1960s, no major technological barriers existed to significantly improve on the Hound Dog in terms of size, range, and probable accuracy. A number of reasons could be advanced, but two seem particularly important. First, from 1965 the war in Indochina became a time-consuming preoccupation for the entire defense community in the United States. The war also became increasingly expensive and was clearly a contributing factor to the precipitous decline in expenditure on strategic forces between FY1964 and FY1967.[14] Second, McNamara had a growing conviction that the United States possessed a veritable glut of strategic capability. Back in 1962, in addition to his analytical objections to the B-70 and RS-70, McNamara stated that the programmed U.S. strategic retaliatory forces "could achieve practically complete destruction of the enemy target system even after absorbing an initial nuclear attack," and that an additional $10 billion for a force of 200 B-70s or 150 RS-70s "would not appreciably change this result."[15] As these programmed forces came to fruition, McNamara became utterly convinced that the United States had overreacted in the strategic arena and he became a vocal champion of restraint, an unusual role for a secretary of defense.

In a sense McNamara, or at least the OSD generally, did attempt to push the Air Force toward long-range, air-launched missiles from 1968 onward, perhaps even from 1967. As mentioned, Air Force bomber penetration studies in 1966-67 had indicated the utility of a successor to the Quail decoy missile. The Quail's most serious deficiency was limited range at low altitude (about 185 km.). It is also reasonable to presume that advances in Soviet radar technology had improved their ability to discriminate between decoys and B-52s. Finally, the Quail's physical dimensions limited the number a bomber could carry to four. In January 1968 SAC started the ball rolling by issuing a required operational capability statement for a new decoy. At about this time, however, the Defense Science Board—a group within OSD—proposed that the new decoy be armed. The issue of whether the new missile should be just a decoy or an armed decoy remained controversial for the next five years until the project was scrapped in June 1973.

The Air Force clearly lost the first round in this battle, in that when the weapon was first unveiled in January 1969 as an integrated project, it was labeled SCAD—subsonic cruise armed decoy.[16] The Air Force's strategy was to make as clear a distinction as possible under the circumstances between the role of SCAD and that of the Hound Dog and the SRAM. In the first public defense of the SCAD program early in 1969, Secretary of the Air Force Seamans argued as follows:

> The Hound Dog . . . is designed to attack . . . targets, such as airfields, which would otherwise require additional low level flight time by the bomber if attacked with a gravity weapon. . . . SRAM . . . is designed to provide the capability for a bomber to either suppress or avoid low altitude capable, terminal SAM defenses. SCAD is designed primarily as a decoy (armed or unarmed) to provide the capability to saturate enemy area defense radars and thus enhance the survivability of the bomber force. Even though the Soviets may be able to discriminate between a bomber and a SCAD, they would be forced to attack all radar objects if only a few SCAD were armed with a nuclear warhead.[17]

This distinction, however, was torpedoed by Director of Defense Research and Engineering John Foster in testimony before the same committee a day or two later. Foster blandly stated that:

> The SCAD carries a nuclear warhead and has adequate range to be used as a stand-off missile so that the bomber does not have to penetrate.[18]

Foster had assumed the post of DDR&E in 1965 and had, at various times, indicated his support for McNamara's view that bombers in the future should basically be launching platforms for missiles.[19] The statement above is perhaps the earliest explicit evidence that an influential group existed within the defense community who believed the technology was available, or could be acquired, to build air-launched missiles that could obviate the requirement for penetrating strategic bombers. This group was not successful in preventing the AMSA proposal from crystallizing into the B-1, a more or less uncompromised penetrating bomber, but it did succeed in preventing the rapid development of SCAD along the lines desired by the Air Force, that is, primarily as a decoy. The Air Force request for $30 million for the SCAD late in 1968 (for FY1970) was cut by OSD to $17.1 million, and Congress reduced it further to $9.1 million. For FY1971 the Air Force and the Department of Defense agreed on a request of $33.6 million, but the testimony on this item gave Congress sufficient cause to deny all of it.

Foster openly acknowledged the dispute between OSD and the Air Force over the relative emphasis to be placed on the armed version of SCAD.[20] The Air Force, in the meantime, had slightly revised its position and was proposing a split SCAD program with SCAD A, a decoy with an optional warhead capability, for the B-52 to be followed by a longer-range SCAD B (also with a warhead option) for the B-1.[21] But the Air Force had not yet finally determined whether SCAD A should be sized so as to fit into the same space as a SRAM missile, and could only offer disquietingly uncertain figures on development costs.[22]

Following the deletion of all funding for FY1971, the Air Force reviewed the SCAD program, decided it was still an urgent requirement, and requested $45 million for FY1972. The OSD, predictably, cut this to $10 million. When the new budget was sent to Congress early in 1971, it was apparent that the seeds planted two years earlier—that aircraft carrying long-range missiles might be an alternative to the penetrating bomber—had begun to take root. Senator McIntyre, for one, wondered why the B-1 development program could not be stretched by two years to "allow a close examination of . . . alternatives including . . . a standoff launch capability employing a nuclear armed decoy."[23] Dr. Jeremy Stone argued a similar case on behalf of the Federation of American Scientists.[24]

Not surprisingly, the Department of Defense, and the Air Force in particular, began to take seriously the evidently rising popularity of long-range, air-launched missiles. It must be remembered that at this time SCAD existed just on paper. The only component that had been demonstrated in any way was the engine, and then only in a very preliminary fashion.[25] Nevertheless, with a design range greater than that of Hound Dog (suggesting a figure of about 1,100 km.), a CEP equal to that of SRAM, and remarkably small overall dimensions, it seemed an extremely elegant weapons system.[26]

The Air Force's main line of counterargument to proposals for a standoff bomber force was the relative vulnerability of bombers and cruise missiles under conditions in which the Soviet Union optimized its air defenses against the respective threats. Air Force officials referred repeatedly to the preliminary results of an in-house study on this subject. The study was based on a calculation that showed that if all the penetrating features were taken off the B-1, its unit cost would be reduced 25 to 30 percent:

> the cost comparisons for forces of equal effectiveness were, in fact, based on having each stand-off bomber cost 30 per cent less than a penetrating bomber. It was for this reason that the cost did swing in favor of the stand-off force when both the penetration and stand-off force were pitted against a Soviet defense that was oriented only against the penetrating bomber. However, it was also noted in the report, if the Soviets reacted to the stand-off threat by simply replacing their SA-3 SAM's with the more sophisticated [deleted] SAM's, the probability of arrival of the force of stand-off cruise missiles is reduced from [deleted] down to [deleted]— while the effectiveness of the penetrating force is not altered by the reactive defense. This occurs because against either defense the penetrating bomber force pays the same price in supersonic SRAM's expended on SAM's in order to reach the targets. On the other hand, the stand-off cruise missile—since it must be subsonic to achieve the necessary range—becomes quite vulnerable to SAM's which have improved capabilities against targets at low altitude. Therefore, to retain an effectiveness for the stand-off force [against the reactive threat] that is equal to that of the penetrating bomber force, the number of stand-off cruise missiles had to be increased—correspondingly increasing the necessary number of stand-off bombers to the extent that the stand-off force would cost more than the penetrating force even though each stand-off bomber was cheaper.[27]

The Air Force further argued that not only would a standoff

bomber force be more costly for the United States, but that defense against this force would be less costly for the Soviet Union. The Air Force estimated that an air defense system optimized against cruise missiles would, over a ten-year period, cost $27 billion less than a system directed against penetrating bombers.[28] The conclusion was almost self-evident: a combination of standoff and penetrating tactics would maximize both the destructive capacity of the U.S. bomber force and the costs of air defense in the Soviet Union.

As shall be seen, the question of the vulnerability of cruise missiles to air defenses became increasingly controversial—and confused; but even at this stage the Air Force firmly believed that "against a subsonic cruise missile in a clear environment [the Soviet SA-3 SAM and follow-on systems] have an excellent capacity."[29] The FY1972 hearings also brought out what might be regarded as the central dilemma of the SCAD system, or indeed of any armed decoy. Viewed as a decoy, the task of the development team was to make a very small vehicle look like a very large B-52 or B-1 on enemy radar screens. To a large extent this is done by putting appropriate electronic devices inside the decoy, but it is also helpful, perhaps even necessary, to shape the decoy so that its inherent radar reflectivity assists the electronic devices as much as possible. Viewed as a weapon, the object would be to maximize the advantages inherent in the vehicle's small size.

Thus Congress found itself listening to statements by Air Force officials that, during the penetration phase of its flight, the survivability of an armed SCAD would be the same as a B-52 because it projected the same electronic image. The Air Force acknowledged that if SCAD operated silently—that is, with the decoy electronics switched off—it could penetrate area defenses (interceptors and long-range, fixed SAMs). But when attacking defended targets the bomber was a far superior penetrator because it could suppress these terminal defenses with hypersonic SRAM missiles.[30] It is of some importance to note that the bomber's advantage was considered to derive only from its ability to carry the SRAM missile. Subsequently a great deal of stress was placed on the inability of cruise missiles to carry electronic countermeasure equipment, a feature that came to be regarded as critical to the viability of the penetrating bomber. The explanation may be that, at this stage (early 1971), even the SCAD was projected to have an active jamming capability in the same frequency bands as the B-52.[31] Whether this capability proved beyond achievement or was dropped for tactical bureaucratic reasons is not known.

In any event, the $10 million requested for SCAD for FY1972 was authorized and appropriated by Congress. The Senate Armed Services Committee had made its support conditional on first priority being given to an increased accuracy, dual-role SCAD, the opposite of what the Air Force had in mind. The full Senate, however, defeated amendments offered by Senator Proxmire to fund studies on two possible standoff bombers: a re-engined B-52 equipped with SCAD and a wide-bodied cruise missile carrier. In the meantime the executive branch had entered the picture in a visible way. In summer 1971 an ad hoc panel of the president's Scientific Advisory Board had concluded that long-range cruise missiles were feasible and would be effective in nuclear war.[32]

In 1972 the Air Force continued to insist that it would develop the SCAD initially as a decoy for the B-52 and defer a more accurate, armed version until this was accomplished.[33] But interest in a decoy was waning rapidly as more people became enthusiastic over the possibilities of attack cruise missiles. The OSD was convinced that cruise missiles could penetrate Soviet air defenses and thus directly contradicted the Air Force position. Commenting on the SLCM, DDR&E Foster stated that "the system would have a very good penetration capability, certainly better than bombers have and as good, maybe better, than that of the SCAD system."[34] In addition, Proxmire came forward with a convincing case against SCAD (as a decoy) in what was perhaps the first case for an ALCM, pure and simple. Proxmire had two objectives: to deny the request to develop a submarine-launched cruise missile and to undermine the rationale for the B-1. In what turned out to be a prophetic argument, Proxmire stated:

> It makes sense to develop a strategic cruise missile as a hedge against threats to our current strategic forces. But the Pentagon is developing the wrong missile at the wrong time. The Navy's SLCM program has little more merit than a nuclear warhead fired from a crossbow, while an attack missile version of SCAD could provide us with an alternative option to the B-1 for preserving our strategic bomber force.[35]

Proxmire went on to state that it was the bomber and not the submarine leg of the Triad that faced penetration difficulties, and that an SLCM program would be contrary to the spirit of SALT I since sea-based strategic forces were limited while bombers were not. He also repeated his charge, first made a year earlier, that the Air Force was deliberately retarding the development of the armed SCAD (for the B-52) so that no hardware would be available at the time a production decision was due to be taken on the B-1. At the time this decision was scheduled for September 1975.

Despite all this the full $48.6 million requested for SCAD for FY1973 was authorized and appropriated and the Air Force apparently felt justified in not volunteering to reorient its declared priorities on the decoy and armed versions. In February 1973, just prior to the congressional hearings on the FY1974 budget, the Government Accounting Office released a study critical of SCAD, pointing out, among other things, that the planned IOC date (1977) was already two years later than the projected threat the system was designed to counter. One can only speculate on what this threat was. The most likely possibility would be the Soviet equivalent of the U.S. AWACS, an aircraft with a large lookdown radar capable of discerning objects flying at low altitudes over land, and with command and control facilities to direct defense forces to intercept these objects. The Soviet Union had deployed such a system in 1970 (the TU-126 Moss), but it only had a lookdown capability over water. United States intelligence may have estimated that 1975 was the earliest date at which the Soviet Union could master the far more demanding overland lookdown capability that would present a serious threat to penetrating bombers.[36]

On March 15, 1973, a DSARC meeting was held to review the program and to decide whether or not to allow SCAD to proceed into engineering development.[37] As the cognizant service, the Air Force presented a paper on SCAD that outlined three alternatives:

A. a decoy with an arming option but no actual development work on the armed version: R and D $285.5 million and procurement $774.4 million;
B. a decoy followed by an armed version: R and D $377.8 million and procurement $826.2 million;
C. a decoy and an armed version with concurrent IOC dates: R and D $405.0 million and procurement $886.7 million.[38]

All these figures excluded the costs of modifying B-52s and developing the warhead that, according to Representative Aspin, would add a minimum of $600 million.[39] The Air Force also acknowledged that the low-level range target for SCAD could not be achieved because of sizing constraints imposed by seeking to fit the weapon on the same launcher used for SRAM. Adding a warhead would further reduce the SCAD's range, but SAC indicated that it would be a usable weapon nonetheless.[40]

The March DSARC meeting did not approve SCAD for engineering development. The Senate Armed Services Committee, angered because the Air Force had ignored its directive to give first priority to the armed version, added to the pressure. The committee's professional staff pointed out that SCAD was matched with an aircraft (the B-52) that would be phased out of service from 1980 onward as the B-1 came in. Unless SCAD was also intended for the B-1, its billion-dollar price tag looked quite excessive. The Air Force, clearly reluctant to associate any additional costs with the B-1, now argued that this aircraft did not require the SCAD. Citing recent studies, Air Force officials stated that a SCAD optimized for the B-1 would add only 0.1 percent to its probability of penetration as against up to 50 percent for the B-52.[41] Moreover, the Air Force asserted that the B-52 was not about to fall apart and the United States could look forward to a mixed force of B-1s and B-52s for some time to come. In information supplied for the record, the Air Force said: "The best available information at this time indicates the service life of the B-52 G/H aircraft to be early to mid-1990's."[42] This information was to have a profound effect on attitudes toward the B-1 and on the amount of effort that could reasonably be devoted to augmenting the capabilities of the B-52.

ENTER THE ALCM

The DSARC group met to reconsider SCAD on 28 June 1973. Shortly afterward the Department of Defense—not the Air Force—canceled the program. Deputy Secretary of Defense William Clements said that $22 million of the $72.2 million requested for SCAD in the FY1974 budget (still under review by Congress) would be used to pursue cruise missile technology, and that this effort would be coordinated with the Navy's SLCM program to avoid duplication.[43]

With the complicating factor of a decoy missile removed from the picture, the debate now began to focus more exclusively on whether air-launched cruise missiles should be viewed as something that enhanced the capabilities of the penetrating bomber force, supplemented it, or replaced it. A second issue that influenced the course of events to a significant degree was Air Force-Navy collaboration and standardization in the cruise missile field.

The Congress as a whole was still skeptical of cruise missiles. In October 1973 the Senate-House Conference on appropriations for FY1974 cut the request for ALCM from $22 million to $11 million, and that for SLCM from $15.2 to $2.5 million. In addition, the conferees insisted that: the monies be used only to develop components and subsystems; the Department of Defense undertake a review of the requirement for cruise missiles and present the results along with the FY1975 budget; and all efforts be made to ensure close cooperation between the Air Force and the Navy.

The unpopularity of the latter directive, at least within the Navy and the Air Force, became apparent very quickly. Early in FY1974 (that is, in the latter months of 1973) the OSD requested the Navy to pursue its SLCM program with the possibility of air-launch from a B-52 in mind.[44] A few months later a Navy official told the House Armed Services Committee that "we are no longer attempting to make our submarine-launched, torpedo-tube launched missile, compatible with launch from an Air Force aircraft."[45] The reason given by the Navy, and supported by the chairman of the Joint Chiefs of Staff, General Brown, was that commonality would unduly restrict the capability of both the air- and submarine-launched weapons.[46] General Brown argued that the Air Force's range requirements were less demanding because the weapon would be carried on a penetrating bomber, and that in any case a weapon that filled a torpedo tube would be too large to fit on the rotary SRAM launcher.

On the role of ALCM, Malcolm Currie, recently appointed DDR&E, was of the opinion that it would "further enhance" the attractiveness of the B-1.[47] Elsewhere in these hearings Currie stated that part of the initial rationale for the ALCM was that the B-1 was a fairly effective cruise missile carrier. The Air Force, consistent with its change of heart on the B-1/SCAD requirement, was somewhat more cautious. Dr. La Berge, the assistant secretary of the Air Force for R&D, was content to say that at some time in the 1980-90 time frame, a cruise missile capable of internal carriage in the B-1 might become a desirable option.[48] Generally speaking, the defense hierarchy endeavored to support the air-launched cruise missile while at the same time blunting the developing tendency to use this weapon as an argument against the B-1. Various officials presented a number of arguments in support of this position. Before these are considered, it is convenient to briefly note the results of the review of cruise missile requirements requested by Congress.

This review, summarized by the Navy for the House Armed Services Committee, pointed out that modern technologies enabled one to overcome the principal deficiencies of earlier cruise missiles (namely, large size and high altitude flight profiles) to achieve adequate range. Most of the other points related specifically to SLCM, but one comment must have whetted the appetites of those who saw in cruise missiles a means of negating the requirement for the B-1. The statement read as follows:

> There is no longer any question that a highly efficient cruise missile, with the capability to penetrate postulated Soviet defenses well into the 1980's, can be built. This has been demonstrated in design studies conducted by both the Air Force and the Navy.[49]

For the Navy, it must be remembered, cruise missiles did not directly threaten established ways of carrying out missions. Although the Navy went to considerable lengths not to aggravate the problems cruise missiles presented for the Air Force, proponents of the B-1 would clearly have preferred a more qualified assessment of their capabilities.

An anonymous Department of Defense response to a prepared question by Senator Hughes maintained that the B-1, given its greater prelaunch survivability, ECM capability, and a crew for decision making, could not be replaced in an overall sense by an air-launched cruise missile.[50] Currie amplified on this by pointing out that a stand-off bomber force would simplify matters for the Soviet Union because they would face a relatively small number of cruise missile carriers that could be attacked either on the ground or outside their borders. In any event, he maintained that the important decisions on cruise missiles would be made much later than the decision to produce the B-1.[51]

Air Force officials pointed out that the range of ALCM was insufficient to enable SAC to strike all the targets assigned to it under the SIOP unless it was carried on a penetrating aircraft.[52] ALCM was regarded as having three functions: to suppress area defenses; attack primary targets not defended by terminal SAM defenses; and to further enhance the unique utility of the bomber within the strategic Triad in limited encounters and shows of force that derived from its high visibility, real-time command and control, and flexible choice of weapons.[53] To emphasize its assertion that the ALCM could not be regarded as a substitute for the B-1, the Air Force flatly stated that it was not even studying the concept of a low-cost platform for the ALCM.[54]

It transpired, however, that there was already some pressure from OSD to do just this. It appears that while the FY1975 budget was being drawn up (late in 1973), "elements of the OSD staff" wanted to add a cruise missile capability to an Air Force proposal for an advanced tanker/cargo aircraft that would be derived from an existing wide-bodied aircraft. The Air Force lobbied successfully to have this additional mission deleted before the budget was presented to Congress, but the Senate Armed Services Committee was informed of the proposal and directed some pointed questions at Air Force officials in consequence.[55]

In pursuing this issue, the point was made that even if some people were exaggerating the ability of cruise missiles to survive Soviet air defenses, the low unit cost of these weapons (the figure at the time was $600,000) could make the saturation of these defenses with large numbers of cruise missiles a cost-effective solution. While not addressing the point directly, the response given by Dr. La Berge is of particular significance in the light of subsequent developments:

> A long-range stand-off cruise missile with any reasonable capability is going to be in the multimillion class, not in

the $600,000 to $800,000 [sic]. Essentially we are subsonic going [deleted] miles, no ECM, we will go low. But to go supersonic, to put ECM in, to do the rest of the things that makes it a viable stand-off weapon, I think you end up talking four or five times as much and saturation becomes an economic question as well as a technical one.[56]

This response was somewhat beside the point. The need to saturate defenses would be far less compelling with the kind of missile La Berge had in mind. Nevertheless, given the pronounced shift a few years later toward viewing the ALCM as a standoff weapon on the ground of cost-effectiveness, his comments are well worth bearing in mind. La Berge further stated that a supersonic cruise missile with a range of 2,000 miles would "be essentially . . . the size of a manned airplane."[57] Moreover, this remark was based on more than armchair analysis. A few years earlier (that is around 1970) the Air Force apparently studied the concept of a supersonic, low altitude penetrating attack missile (LAPAM) and came up with an unacceptably large and heavy weapon.[58] In additional comments, the Air Force stressed the inherent incompatibility in cruise missiles between small size and low radar cross section on the one hand and high speed and long range on the other. It acknowledged that ALCM, by virtue of its low penetration altitude, could survive existing Soviet defenses. But against projected defenses in the 1980s, which comprised AWACS, lookdown/shoot-down interceptors, and better low altitude SAMs, a true standoff cruise missile would need a very low RCS <u>and</u> a speed of Mach 3 plus.[59] A vehicle with this speed and range capability could not be built with an inherently small RCS, so it would have to carry ECM equipment in order to disguise itself and probably an automatic electronic counter-countermeasure (ECCM) capability because, if detected, it could not alter speed or maneuver in order to protect itself.[60]

From this point on the ALCM issue was profoundly influenced by two major studies on the future of the strategic bomber force. The first was the Joint Strategic Bomber Study (JSBS) prepared by the Air Force and the Department of Defense and delivered to Congress in December 1974. The second was prepared by two analysts from the Brookings Institution and appeared approximately a year later, although Congress was informed of its preliminary results early in 1975.[61] The studies arrived at opposing conclusions: the JSBS demonstrated the cost-effectiveness of the B-1, while the Brookings study suggested the B-1 was unnecessary for the time being and that when a new bomber became necessary, it would not have to be as sophisticated as the B-1.

Armed with the JSBS, the Air Force faced Congress again in the early months of 1975 in hearings on the FY1976 defense budget.[62]

AIR-LAUNCHED CRUISE MISSILE / 75

There was little change in the line of argument. Given the projected threat in the 1980s and beyond, the B-1 was indispensable while the ALCM would substantially augment the capabilities of the B-52s and, indeed, would become an essential weapon for this aircraft it it were to be retained as an operational system into the late 1980s and early 1990s. One official summarized the position as follows:

> if the target is defended with an advanced low-altitude SAM, the cruise missile is not expected to be able to penetrate the defense. . . . it would not be cost-effective to saturate or exhaust such a defense with cruise missiles . . . because such an advanced SAM defense would be very expensive, there are many targets in the Soviet Union which probably will not be defended with these types of SAM's. Against undefended targets, cruise missiles are cost effective compared to B-1s and B-52s carrying SRAMs or gravity bombs.[63]

If one were inclined to be cynical, one might detect a new tendency on the part of the Air Force to protect the B-1 by hinting that it would be necessary to develop a number of additional weapons for the B-52 if the B-1 were canceled or postponed. Thus, in addition to the ALCM being "crucial" for the B-52 as against merely "advantageous" for the B-1, the Air Force argued that if the B-52 were retained on its own, there was a much higher probability that it would be necessary to develop and procure a short-range bomber defense missile and the advanced strategic air-launched missile that, in one of its roles, would be a long-range bomber defense missile.[64] Both these weapons were in the early design stage at the time.

Insofar as delays in the ALCM program would make a favorable decision on the B-1 more likely, the Air Force received some assistance from the OSD. Whereas during the preparation of the FY1975 budget the OSD had sought to accelerate the ALCM program by nearly doubling the funds requested by the Air Force, it moved in the opposite direction on the FY1976 budget. Presumably in response to congressional concern over Air Force/Navy duplication in the cruise missile area, the OSD decided to keep the ALCM in advanced development and to synchronize this program with the Navy's SLCM, which was at a less advanced stage of development. In fact, it appears the ALCM had a brush with death in that some consideration was given to its termination in favor of an air-launched version of SLCM.[65] Both weapons were retained but, as before, the Navy was asked to keep the B-52 in mind as a potential launch platform for SLCM.

A related issue that lingered on during these hearings was the Air Force's determination to constrain the dimensions of the ALCM so that it would be interchangeable with SRAM. This limited the weapon's range to 700 or 750 miles, quite inadequate for the standoff role. The Air Force was examining a longer-range version but its interest was clearly marginal; what it had in mind was a cumbersome version of the ALCM with a jettisonable fuel tank slung underneath. Such a weapon could not have been carried internally by either the B-52 or the B-1.[66]

A final issue that must be mentioned is the initial response to the agreement between Brezhnev and Ford, reached at Vladivostok in November 1974, on guidelines for future SALT negotiations. The principal features of the new agreement were a limit of 2,400 on the overall number of strategic delivery vehicles and a sublimit of 1,320 on the number of ICBMs and SLBMs equipped with MIRV. The short published text of the agreement seemed quite specific that the MIRV limit applied to numbers of ICBMs and SLBMs, but there was apparently some doubt even at this early stage as to whether or not bombers armed with missiles should be regarded as MIRVed systems. In addition, the Vladivostok accord covered "certain long-range air-to-surface missiles."[67] It transpired subsequently that the United States and the Soviet Union differed on what had been agreed at Vladivostok on this issue. The Air Force presumed that ALCM was exempt. As Secretary of the Air Force McLucus put it: "We are continuing development. Someone would have told us if there was an effect."[68]

For the various long-range cruise missile programs, 1976 proved to be an eventful year: a production decision on the B-1 was scheduled for November and both opponents and proponents were now backed by major cost-effectiveness studies; the DDR&E undertook a detailed examination of the viability of replacing the ALCM with an air-launched version of SLCM; it was a presidential election year and the issues of U.S.-Soviet relations, detente, SALT, and the strategic balance were prominent; and the ALCM completed its first successful bomber-launch test in March.[69]

The debate on the relative merits of the JSBS and the Brookings study on the B-1 versus cruise missile options for modernizing the strategic bomber force was at best inconclusive. The two studies were built on fundamentally different assumptions, particularly regarding the role of strategic bombers and on the extent to which bombers could presume support from the other two legs of the Triad in fulfilling their role. To a large extent, therefore, the confrontations did not take place on common ground, and the outcomes could usually be regarded as satisfactory by both sides.

Although the two studies differed on numerous points regarding methodology and costing, the Air Force rested its case predominantly on two points: the air-launched cruise missile lacked the range to reach all the targets assigned to the bomber force under the SIOP; and it had an unacceptably low probability of penetrating to targets protected by low altitude SAMs. The estimated maximum range of ALCM was still approximately 1,300 kilometers at penetrating altitudes (200 feet or less), but its effective range would be degraded: if it flew an indirect course in order to avoid SAMs or high terrain (to reduce the chances of detection); and if it were carried on a nonpenetrating aircraft and therefore had to be launched from outside the Soviet defense perimeter usually taken as a nominal 320 kilometers (200 miles). A further important constraint on the utility of ALCM (or indeed any of the U.S. long-range cruise missiles) was the availability of sufficiently varied terrain along the preferred flight path and particularly at a point very close to the target to permit reliable use of the TERCOM guidance system.

Accordingly, in the Air Force's view, there existed a core of targets—those deep inside the Soviet Union and/or those heavily defended with SAMs—against which the B-1 was most cost-effective, while B-52-delivered ALCMs could efficiently tackle the remainder.[70] Having established this case, the Air Force was unreservedly enthusiastic about cruise missiles as the following statement by Thomas Reed, the new secretary of the Air Force, makes clear:

> The air-launched cruise missile . . . has a technology with tremendous growth potential, and we see embarking on technology work that is going to lead to a second, third and fourth generation air-launched cruise missile which 20 years from now will have a tremendous impact on how we maintain our national security.[71]

The potential spanner in the works was a SALT II agreement within the guidelines established at Vladivostok. The chief of staff of the Air Force, General Jones, was sufficiently impressed with the ALCM to suggest that if bombers armed with ALCMs were counted under the limit of 1,320 MIRVed delivery systems, "then it might be desirable to cut a limited number of MIRVed missiles. . . . If a large number of carriers were to be counted as MIRVs (for example, the entire B-52 G and H force) then it would not be desirable to proceed with ALCM deployment."[72] The first part of this statement referred to the fact that, at the time, the number of existing and firmly planned MIRVed missiles in the U.S. arsenal was 1,286, leaving just 34 for bombers armed with ALCM. Such a limited deployment was presumably considered not to be cost-effective. The latter part of the statement presumably referred to the possibility that verification considerations would preclude having some B-52s counted as MIRVed systems and others not. Counting them all as MIRVs if some carried ALCMs would parallel the solution suggested for the verification difficulties presented by MIRVed missiles, namely, that any class of missile tested with MIRV would be assumed to be MIRVed when deployed.

On June 30, 1977, President Carter announced his stunning decision not to proceed with production and deployment of the B-1. The decision was rationalized predominantly on the ground of cost-effectiveness (the details of this rationalization will be examined below). Also influential was Carter's ambition to balance the federal budget by the end of his first term: a production decision in mid-1977 would have led to annual expenditures on the program reaching very high levels by 1980. The B-1 decision, however, was preceded by several important developments in the ALCM program that, with the benefit of hindsight, gave a strong indication of which way the decision would go.

In the FY1978 defense budget drawn up by the Ford administration, the rough estimate of the number of ALCMs to be procured (the figure used in cost calculations) was increased more than threefold. The old figure of 700 missiles was based on the assumption that only B-52s would be armed with ALCMs, and that only a small proportion of the B-52 fleet would be suitably modified. The new figure of 2,328 missiles was based in part on modifying additional B-52s to carry ALCMs, with the target being to eventually modify 15 squadrons or over 200 aircraft. However, the greater part of the increase in planned ALCM procurement stemmed from the decision to include the B-1 as a launch platform.[73] It will be recalled that, previously, the Air Force had regarded the B-1/ALCM combination as strictly an option for the future. It is not clear whether the Air Force changed its mind or whether it was overruled by the OSD, but the latter seems more likely.

A DSARC meeting in January 1977 resulted in a particularly significant restructuring of the ALCM program. First of all, it was decided to completely formalize Air Force-Navy cooperation in the cruise missile field by creating a joint project office. To the Air Force's chagrin the Navy was put in charge.[74] The really significant decision was to accelerate the development of the extended-range ALCM (ALCM-B) so that it would be available earlier than the ALCM-A, an IOC in December 1980 as against July 1981 for the latter. As seen previously, the Air Force had no more than a marginal interest in any version of the ALCM other than the one that could be fitted directly onto the internal rotary launcher for the SRAM. It has also been seen that many other officials, including senators and congressmen, regarded this preoccupation with ALCM-A as a move to protect the notion of a penetrating bomber. This latter view clearly carried the day at the DSARC II meeting. General Slay, a deputy chief of staff of the Air Force, provided the following assessment:

> Rightly or wrongly the impression that had been left with the DSARC was that the Air Force was much more inter-

ested in carrying the smaller ALCM internally in a penetrating mode. I guess the thought was that we should bring the longer-range version on at the same pace . . . so that we would have a standoff missile in our inventory.[75]

Nevertheless, the ALCM-A remained an active program and, according to another official, the DSARC group still supported the view that cruise missiles could not be depended upon to penetrate SAM-defended targets so that penetrating bombers carrying SRAM would still be required.[76] While this may have been of some comfort to the Air Force, another aspect of the decision to accelerate ALCM-B was filled with ominous potential.

The DSARC group determined that ALCM-B would be an elongated or stretched version of the ALCM-A rather than what the Air Force had had in mind since 1973: an ALCM-A with an extra fuel tank slung underneath. Coupled with the new preference for ALCM-B, this decision virtually had the effect of ruling the B-1 out of the long-range cruise missile stakes. The reason was that the B-1 had three short bomb bays sized to accommodate SRAM and gravity bombs. Anything significantly longer simply would not fit, and the proposed ALCM-B was five feet longer than SRAM.

Once the decision had been made to give precedence to a longer-range ALCM, a preference for the elongated version of ALCM-A was inevitable. The version with an external fuel tank could not be carried internally by either the B-1 or the B-52, and it had less range than the stretched version, 1,250 n.m. versus about 1,400 n.m.[77] A B-52 would be able to carry a maximum of 20 stretched ALCMs: 8 internally on a redesigned rotary launcher and 12 in two clusters of 6 on external pylons. Significantly, it was stated at the time that the internal carriage of ALCM-B in the B-52 would preclude the carriage of any other weapon internally, be it SRAM, ALCM-A, or gravity bombs.[78]

On the question of a common missile for all the proposed launch platforms, the DSARC II group accepted the view that this might impose unnecessary and unwarranted performance compromises on the ALCM and SLCM. Both missiles were therefore approved for full-scale engineering development. It can be pointed out, however, that the willingness to accept the penalty of having to develop a new launcher for the ALCM-B greatly improved the prospects for the air-launched version of the SLCM.

This was the position when the Carter administration assumed office at the end of January 1977. A change of administrations is not abrupt. There is a transition period between the elections in November and the inauguration in January during which the new team works with the old. Major policy initiatives are avoided and controversial issues tend to be suspended. Thus a production decision on the B-1 was postponed to permit a review by the new administration. Similarly, the new team presumably had some influence in the January DSARC meeting on cruise missiles. Generally, however, the Carter administration's flexibility in the defense field was limited by the fact that consideration of the FY1978 budget was well advanced. Many amendments were proposed but these consisted predominantly of small reductions in or a rescheduling of particular programs. A salient example of the latter, eventually accepted by Congress, was the transfer of $24 million from the tactical antishipping version of SLCM to the GLCM account, which then totaled $27.9 million.

President Carter's secretary of defense, Harold Brown, promptly initiated a thorough reexamination of the cost-effectiveness of alternative bomber forces. At the technical level this reexamination was conducted by Robert Perry, successor to Malcolm Currie as DDR&E. While the Air Force was undoubtedly nervous (and with good reason) about the prospects for the B-1, it probably drew some comfort from the fact that the secretary of defense was a past secretary of the Air Force, and in that capacity had been a firm if not ardent supporter of a follow-on to the B-52.[79]

The new study took three months to complete. Perry then reported the results to Brown who, in turn, briefed the president. Carter apparently spent several weeks going through mountains of background material and eventually made his decision in virtually complete isolation. Even the chairman of the Joint Chiefs of Staff was informed of the decision only on the morning of the day it was to be announced. Carter made it clear that his decision was not so much against the B-1 as it was in favor of cruise missiles launched from B-52s (and possibly, at some future date, wide-bodied aircraft) as the most cost-effective way of preserving the viability of the bomber leg of the strategic Triad.

Clearly, the results of the JSBS had been overturned.[80] Broadly speaking, changes in assumptions or judgments in three main areas could have produced this result:

(a) A change in attitude toward risk taking in the strategic arena, particularly in the form of the assumption that Soviet air defenses would be suppressed by ballistic missile attack prior to the arrival of the bombers;

(b) A significant relative change in the calculated capabilities of penetrating and standoff bombers to deliver weapons on targets; and

(c) A change in projections of the magnitude and sophistication of the air defense systems that the Soviet Union would be able to develop and deploy in the future.

It became clear subsequently that the two related factors (b) and (c) were particularly influential in the decision to cancel the B-1. In the first place, by March 1977 there had been a year of testing of full-scale R&D models of the cruise missile, whereas the JSBS worked with a far more limited data base on these weapons. This testing experience permitted more confident projections of the abilities of cruise missiles to penetrate SAM defenses. Even more important, however, was the reevaluation of the potential threat. The JSBS assumed the Soviet Union could deploy defenses that would be highly lethal to cruise missiles by the time the United States could deploy a substantial force of these weapons (around 1985). The reassessment concluded that the defenses the Soviet Union could actually field by that time would be far less effective, so that the attrition rate for cruise missiles computed in the JSBS was substantially revised.[81]

Jointly, these changes had two ramifications. First, cruise missiles deployed on existing platforms (the B-52s) came out as more cost-effective than a force of B-1s when the time frame of comparison was limited to the 1985-90 period. Second, the reduction in the projected severity of the threat allowed the foreseeable capabilities of second- and third-generation cruise missiles to assume more prominence in that time would be available to develop and incorporate such improvements in cruise missiles as improvements in Soviet defenses dictated. Just as the B-1 had been touted as the bomber weapons system most insensitive to changes in the character of the threat, the new judgment was that "we have come to appreciate that cruise missiles can be improved faster than the Soviets could upgrade their SAM defenses."[82] The secretary of defense encapsulated this last notion as follows:

> I have more confidence in our estimates of the effect that the low detectability of the cruise missile will have on Soviet radar than in the effect that the B-1's radar countermeasures would have had.[83]

The reduction in the projected severity of the threat also extended the time period over which suitably modernized B-52s would remain viable as penetrators. Thus while the proposed bomber force would consist predominantly of B-52s carrying long-range cruise missiles, some late-model B-52s and all the FB-111s would remain equipped with SRAM and gravity bombs and would penetrate into the Soviet Union. The rationale for this mixed force was familiar: some targets warranted the larger nuclear yields available in bombs, and it would prevent the Soviet Union from optimizing its air defenses against cruise missiles.[84]

The acquisition cost of the proposed B-52/ALCM force was estimated to be just 27 percent of that of the original B-1 force of 244 aircraft (Table 4.1).[85] This, however, is an unfair comparison because

TABLE 4.1

B-1 and B-52/ALCM Program Acquisition Costs
(then-year dollars, billion)

	B-1	B-52/ALCM
RDT&E	4.2	0.7
Procurement	20.6	3.3
B-52 modification/modernization	—	2.8
Total	24.8	6.8

Source: Hearings on HR 8390 and Review of the State of U.S. Strategic Forces, House Armed Services Committee, July-November 1977, p. 83.

the full B-1 force would apparently have put more weapons on target. Thus the new administration rejected the notion that the United States needed the full capabilities of the B-1 force. Instead, it specified the task required of the bomber force and concluded that:

> a force of [deleted] unit equipment B-1's cost about $13.6 billion and an equally effective force of [deleted] B-52's with 20 cruise missiles each costs about $9.6 billion. These are 20 year program costs (discounted at an annual rate of 10 per cent) which include acquisition but exclude costs prior to 1978.[86]

The B-1 decision was followed up with an amendment to the FY1978 budget requesting an additional $325.4 million for cruise missiles.[87] The principal features of this amendment were: an additional $50 million for ALCM-B development to bring its IOC forward from December 1980 to March 1970 (ALCM-A would be canceled); $103 million for the air-launched version of SLCM to enable this weapon to be available for a competition with the ALCM-B starting in January or February 1979; an additional $60 million for preparations for the production of cruise missiles; and $90 million to investigate the concept of a wide-bodied cruise missile carrier.[88]

This program was expected to yield a B-52/ALCM force on about the same schedule as had been established for the B-1. The number of ALCMs to be procured was increased to 3,424 to arm the 173 B-52G bombers.

The B-1 decision was so unexpected that, for a time, comment and debate tended to overlook the fact that the Carter administration had not abandoned the penetrating bomber, neither for the immediate future nor as an option for the more distant future. Stated baldly, the Carter review of bomber modernization concluded that the Soviet Union could not, over the next ten years or so, field air defense systems that could prevent a mixed force of penetrating and standoff B-52s from putting nuclear weapons on 80 percent of the Soviet target system.[89] A decision to proceed with the B-1 would have represented an excess of capability, at least through the 1980s. Moreover, this excess of capability carried a handsome price tage. If the full capabilities of the projected B-1 force were quite unlikely to be required until about 1990, it made good sense not to proceed with production in 1977. For one thing, Soviet air defense technologies could evolve in a way that would render the particular mix of capabilities embodied in the B-1 less than optimal. For another, technological advances in offensive and defensive bomber weapons and equipments (particularly the cruise missile), in engines, and so on, would present new opportunities to the designer of a strategic bomber in the future.

Although Carter's recommendation on the B-1 was accepted by Congress, his proposed B-52/ALCM alternative was naturally subjected to particularly intense scrutiny. One obvious ramification was that the administration's negotiating position on cruise missiles at SALT was now far more constrained. At the press conference announcing the proposal to cancel the B-1, Carter stated that the prevailing U.S. position was that bombers equipped with cruise missiles should not be considered MIRVed systems, although he indicated that this position was negotiable.[90] On the question of range, it was widely accepted that the minimum for a true standoff missile would be 2,500 km. In fact, since most cruise missiles would be programmed to fly indirectly to their targets, an operational range of 2,500 km. would require a missile capable of flying somewhat farther. Thus DDR&E Perry stated: "we anticipate that cruise missile range limitations will apply to the operational systems range."[91]

The major part of the debate centered on the possibility that the administration had underestimated the speed with which the Soviet Union could pull together the technologies needed to negate the first-generation subsonic cruise missiles. One issue was how rapidly the Soviet Union could extend the standoff distance necessary to ensure the survivability of the launching aircraft. The notional figure of 200 n.m. used in the past was now contradicted by Defense Department studies that indicated that distances of 500 to 800 n.m. would be more realistic in the near future.[92] If the carriers approached at high altitude, they could be detected well away from the Soviet Union's borders by existing surveillance and detection systems and could be attacked by long-range interceptors.[93] The Air Force's preference was for a cruise missile capable of reaching 87 percent of the counterforce targets (that is, military targets) in the Soviet Union.[94] The deepest of these targets are some 2,300 km. from the coastline. A SALT II limit of 2,500 km. on ALCM range would put these targets out of reach since the balance of 200 km. is insufficient to permit launch from a safe standoff distance and to permit the weapon some evasive maneuvers en route to the target.[95] At this time only the Tomahawk had the range (about 3,800 km.) to permit the desired degree of target coverage even when launched 800 n.m. (1,480 km.) off the coastline.

The official position on this question was that as long as the B-52 remained the primary cruise missile carrier, the option existed to bring it down to low altitude from several hundred miles prior to the launching of cruise missiles. In this event, a credible Soviet threat to the carriers would have to consist of a fleet of AWACS aircraft and an interceptor force with lookdown/shoot-down capabilities. Since there was no evidence that the Soviet Union possessed either of these aircraft types in 1977, it was felt that the time required to develop and deploy them in the quantities required extended well beyond the anticipated life span of the first-generation cruise missiles.[96]

The second issue concerned the vulnerability of the cruise missile en route to its target. The official attitude on this issue has already been described: namely, that the low penetrating altitude and small radar cross section of the cruise missile would defeat existing Soviet defenses and its capabilities could be improved faster than the Soviets could upgrade their defenses. In the months following the B-1 decision, a campaign was mounted to undermine the credibility of this assessment. The counterargument rested on the existence of a new SAM, the SA-10, and the appearance in the Soviet Union of radars mounted on towers several hundred feet high to permit the detection of low-flying objects at longer ranges.

The radar towers, observed under construction along the Soviet borders with Poland and Romania and first made public in July 1977, are probably directed at the low-level threat in general, that is, NATO's tactical nuclear aircraft as well as strategic systems like the B-52 and B-1. As far as the cruise missile is concerned, NATO officials were reportedly unimpressed with the new countermeasures.[97] The radar cross section of ALCM or SLCM is estimated to be less than 15 percent that of an F-4 Phantom fighter-bomber and 0.2 percent that of a B-1.[98] Other commentators, however, presented the new radar network as the first link in an effective Soviet defense against cruise missiles.[99] By December 1977 reports emerged that the Soviet Union was testing mobile, tower-mounted radars. Unlike the fixed towers just mentioned, these were specifically designed to counter the cruise missile.[100]

The SA-10 issue was far more controversial and provides a good example of how patchy intelligence information (or, in other words, secrecy) can generate an atmosphere of urgency. In July 1977, just after the B-1 decision, <u>Aviation Week and Space Technology</u> reported that the Soviet Union had developed a hypersonic SAM to use against the cruise missile.[101] The existence of this missile was not officially acknowledged until October. In that month two journalists, working on the basis of information supplied confidentially by sources in the Pentagon, claimed that recent computer studies had demonstrated the vulnerability of U.S. cruise missiles even to current-generation SAMs, let alone the "more advanced . . . SAM-10 now guarding the Soviet homeland."[102] Official spokesmen denied the existence of these computer studies and claimed the operational deployment of the SA-10 was some eight years away.[103] But in December <u>Aviation Week and Space Technology</u> also claimed that the new SAM was already deployed.[104]

Curiously, the posture statement for FY1979, released in January 1978 by the chairman of the Joint Chiefs of Staff (and typically the most conservative official document cataloging the state and trend of the military balance), made no mention of the SA-10. To the contrary, it suggested that the effectiveness of the 6,000 radars, 2,600 fighter interceptors, and 12,000 strategic SAMs that made up the Soviet air defense system "against low level tactics . . . is less than these sheer numbers would suggest."[105] In March 1978, positions on the character and status of the SA-10 diverged even further. <u>Aviation Week and Space Technology</u> reported that the weapon had been moved from the Soviet SAM development complex near the Chinese border in the east to a test site in the west, and suggested that deployment early in 1979 was likely. It also described the weapon as having a speed in excess of Mach 6, an active radar terminal seeker, and a range of 27 nautical miles.[106] A few months later the official position, as stated by DDR&E Perry, was still that the SA-10 was under development, not deployed.[107]

This mini-controversy on the status of the SA-10 was related to a planned series of tests in which the Navy's SLCM would be flown against U.S. SAMs and simulated Soviet SAMs. The results of the first series of tests, held over the period January-October 1978, were most reassuring. Perry stated that existing Soviet air defenses would be "totally ineffective" against the ALCM.[108] It was acknowledged that the Soviet Union had (or would soon have) the technologies needed to mount an effective defense, but the numbers of systems required—50 to 100 AWACS-type aircraft, 500 to 1,000 SA-10 complexes, and several thousand interceptors with a lookdown radar capability—would absorb $30 to $50 billion and take about 10 years to deploy. This would afford the United States ample time to field a second-generation ALCM capable of defeating these new defenses.[109]

In the midst of this debate, Defense Secretary Brown presented his first annual report in February 1978. In it he reaffirmed the Carter administration's commitment to the long-range, air-launched cruise missile. Indeed, he stated that this weapon's program "now has our highest national priority."[110] The decision in favor of the B-52/ALCM force was explained in terms almost identical to those described above but, in addition to cost-effectiveness, several other advantages were put forward. Specifically, the B-52/ALCM force would: curb the "current trend toward excessive reliance on SLBMs"[111]; "improve the world's perceptions of the potency of [U.S.] forces"[112]; and represent a major upgrading of the strategic forces without threatening a first-strike capability.[113]

The long-range strategic cruise missile had clearly come into its own although deployed on aircraft rather than submarines as originally envisaged. By the end of 1977 there was no longer any question that the ALCM program would represent the greater part of the modernization of the bomber leg of the strategic Triad over the ensuing decade or more. It was also strongly implied in the proposal for a wide-bodied cruise missile carrier and in general statements to the effect that cruise missile technology was in its infancy, that these weapons would become increasingly prominent in the U.S. military posture over time. All this was borne out in the military budget for FY1980 submitted to Congress in January 1979. The sum requested for the AGM-86B totaled $475.4 million, of which $90 million was for R&D and the remainder for the procurement of 225 missiles, initial spares, and associated construction.[114] The wide-bodied cruise missile carrier, an advanced cruise missile technology program, and the ASALM accounted for an additional $65 million.[115] Significantly, there was also a $5 million request for research on a new manned strategic bomber.[116] The point was made above that, in canceling the B-1, the Carter administration did not argue that the days of the penetrating strategic bomber were over, only that it was cost-effective to have a mixed penetrating/standoff bomber force and that future manned bombers should be designed with both roles in mind.[117]

The vicissitudes of the ALCM program after the decision to cancel the B-1 are not particularly relevant to this study. That decision established the commitment to develop and deploy a long-range, air-launched cruise missile for strategic use. The Carter administration did not waver from this commitment. The flyoff competition was conducted in the latter months of 1979, and in March 1980 Boeing's AGM-86B was declared the winner. By that time Boeing's missile had been lengthened somewhat to improve range, and its unusual duckbill nose—a design feature originally dictated by the position of the gearbox on the SRAM rotary launcher—had given way to a more aerodynamic cone shape. The procurement program remains fixed at 3,418 missiles. The first B-52 configured to carry 12 missiles externally is planned to be available in September 1981.

NOTES

1. Air Force Magazine, May 1973, p. 151.
2. Chairman of the Joint Chiefs of Staff, General George S. Brown, United States Military Posture for FY1979, p. 32.
3. This activity, using one's own defensive capabilities to assess the viability of offensive weapons, is used by some to suggest that the United States, with a wide margin of technological superiority, is really racing itself rather than the Soviet Union. Two observations can be made in this context. First, because the Soviet Union is an intensely secretive opponent there is often little alternative, and second, SAM technology is one area in which the Soviet Union has always matched if not outperformed the United States, particularly for land-based systems.
4. Capt. J. B. Wooley in "Final Evaluation Report on GAM-67 (Crossbow) (System 121A," 12 August 1958. Document read at the Albert F. Simpson Historical Research Center of the United States Air Force, Maxwell Air Force Base, Alabama.
5. Untitled document read at the Albert F. Simpson Historical Research Center in Alabama.
6. Missiles and Rockets, 28 July 1958, p. 109.
7. John W. R. Taylor, ed., Jane's All the World's Aircraft 1960/61 (London: Sampson, Low, Marston, 1960), p. 463.
8. Missiles and Rockets, 28 July 1958, p. 111. At one point in the late 1950s Paul Nitze was advocating an armed mobile, ground-launched variant of the Quail in preference to the Thor and Jupiter IRBMs then being developed for deployment in Europe. He regarded the latter weapons as destabilizing because they would be highly threatening to the Soviet Union and vulnerable to attack.
9. Bill Gunstan, Bombers of the West (London: Ian Allen, 1973), pp. 252-53.
10. Ibid., p. 255.
11. Quoted in ibid., p. 256.
12. Status of U.S. Strategic Power, hearings, Preparedness Investigating Subcommittee of the Senate Armed Services Committee, April 1968, p. 134.
13. Military Procurement Authorization for Fiscal Year 1968, hearings, Committee on Armed Services and the Subcommittee on the Department of Defense of the Committee on Appropriations, Senate, January-February 1967, p. 287. In March 1967 the House Armed Services Committee more or less openly accused McNamara of prejudice against the manned bomber (Hearings on Military Posture and a Bill [HR9240], House Armed Services Committee, March-April 1967, pp. 478-94).
14. Albert Wohlstetter, "Rivals, But No Race," Foreign Policy (Fall 1975): 62. Shortly after McNamara's departure in February 1968

88 / ORIGIN OF THE STRATEGIC CRUISE MISSILE

the new secretary of defense, Clark Clifford, cited war costs as one reason for deferring the full-scale development of AMSA (Hearings on Military Posture and an Act [S.3293], House Armed Services Committee, April–June 1968, p. 8,617).

15. Aviation Week and Space Technology, 19 March 1962, p. 38.

16. The first concrete indication that a new project was under way came in June 1968 when money was provided for preliminary study of the electronics package for a new decoy (Space Business Daily, 7 October 1973).

17. Authorization for Military Procurement, Research and Development, Fiscal Year 1970 and Reserve Strength, hearings, Senate Armed Services Committee, March–April 1969, p. 1,057.

18. Ibid., p. 1,846.

19. See, for example, Status of U.S. Strategic Power, hearings, p. 55.

20. Authorization for Military Procurement, Research and Development, Fiscal Year 1971 and Reserve Strength, hearings, Senate Armed Services Committee, February–March 1970, p. 352.

21. Hearings on Military Posture and Legislation, House Armed Services Committee, February–April 1970, p. 8,225.

22. The estimate provided for SCAD A was $210 million, but the assistant secretary of the Air Force for R&D pointed out that key performance parameters for the missile (range, accuracy, and ECM capability) were not yet resolved. He indicated that the final figure, including inflation, would probably top $400 million (Authorization for Military Procurement, Research and Development, Fiscal Year 1971 and Reserve Strength, hearings, p. 1,179).

23. Fiscal Year 1972 Authorization for Military Procurement, hearings, Senate Armed Services Committee, March–May 1971, p. 492.

24. Hearings on Military Posture and HR3818 and HR8687, House Armed Services Committee, March–May 1971, pp. 2,791–95.

25. In 1968 the Williams Research Corporation was provided with $1.65 million to improve the thrust and efficiency of a turbofan engine that had previously been briefly considered by the Army as a man-pack propulsion system. Testing of this engine in November 1970 had indicated the feasibility of acquiring small, high-performance turbofans.

26. The accuracy objective was given in Fiscal Year 1972 Authorization for Military Procurement, hearings, p. 3,106. The accuracy of SRAM is reportedly very good, and indeed its terminal-defense-suppression role would require it. It is likely that this ambitious accuracy objective was established by those who preferred the armed SCAD, that is, the OSD. The Air Force indicated a willingness to settle for less accuracy and argued that this could hold down development costs.

27. *Fiscal Year 1972 Authorization for Military Procurement*, hearings, pp. 2,873-74.
28. Ibid., pp. 2,876-77.
29. Ibid., pp. 2,881-82.
30. Ibid., p. 3,134.
31. Ibid., p. 3,083.
32. It seems that the panel only endorsed air-launched cruise missiles. This can be deduced from the fact that one member, Richard Garwin, was later highly critical of the proposal to develop a strategic submarine-launched cruise missile. See Garwin's testimony in *Strategic Arms Limitation Agreements*, hearings, Senate Foreign Relations Committee, June-July 1972, pp. 355-56.
33. *Hearings on Military Posture and HR12604*, House Armed Services Committee, February-March 1972, p. 10,928.
34. *Fiscal Year 1973 Authorization for Military Procurement, Addendum No. 1, Amended Military Authorization Request Related to Strategic Arms Limitation Agreement*, hearings, Senate Armed Services Committee, June-July 1972, p. 4,353.
35. Quoted in *Space Business Daily*, 6 July 1972, pp. 18-19. Proxmire's hyperbole on the SLCM was inspired by Paul Warnke, who used much the same language in testimony before the Senate Foreign Relations Committee. Five years later Warnke became the chief U.S. negotiator to SALT.
36. If this was the case, it was an optimistic forecast (or pessimistic depending on one's point of view); the first indications that the Soviet Union was developing an overland AWACS did not appear until late 1978.
37. DSARC stands for Defense Select Acquisition Review Council. All major weapon systems go through the DSARC process at least twice; once before going into engineering development (the production and testing of prototypes) and again before being authorized for production.
38. *Hearings on Cost Escalation in Defense Procurement Contracts and Military Posture and HR6722*, House Armed Services Committee, April 1973, p. 3,199.
39. *Space Business Daily*, 29 March 1973, p. 163.
40. *Fiscal Year 1974 Authorization for Military Procurement*, hearings, Senate Armed Services Committee, March-May 1973, p. 3,354.
41. Ibid., p. 1,139.
42. *Hearings on Military Posture and HR6722*, p. 3,188.
43. *Space Business Daily*, 10 July 1973, p. 38.
44. *Hearings on Military Posture and HR12564*, House Armed Services Committee, February-April 1974, p. 4,148.
45. Ibid.

46. Fiscal Year 1975 Authorization for Military Procurement, hearings, Senate Armed Services Committee, February-April 1974, p. 391.

47. Hearings on Military Posture and HR12564, p. 3,518.

48. Ibid., p. 4,271.

49. Ibid., p. 4,149.

50. Fiscal Year 1975 Authorization for Military Procurement, hearings, p. 286.

51. Ibid., pp. 900-1, 2,074.

52. Ibid., p. 2,913.

53. Ibid., p. 419.

54. Ibid., p. 2,913.

55. Ibid., pp. 2,724-26, 2,893.

56. Ibid., p. 3,921.

57. Ibid., p. 3,938.

58. Ibid., p. 3,939.

59. At the time a U.S. AWACS was imminent and lookdown/shoot-down interceptors were just becoming operational, that is, the projected threat for the 1980s was not unrealistic.

60. Fiscal Year 1975 Authorization for Military Procurement, hearings, p. 3,939.

61. Alton H. Quanbeck and Archie L. Wood, Modernizing the Strategic Bomber Force: Why and How (Washington, D.C.: Brookings Institution, 1976).

62. The U.S. government was moving to a new fiscal year starting on 1 October 1976 so that the FY1976 budget covered the 15 months from 1 July 1975 to 30 September 1976.

63. Hearings on Military Posture and HR3689, House Armed Services Committee, February-May 1975, p. 1,676.

64. Fiscal Year 1976 and July-September 1976 Transitional Period Authorization for Military Procurement, hearings, Senate Armed Services Committee, Part 4, February-March 1975, p. 1,995.

65. Ibid., p. 2,011.

66. Ibid.

67. Ibid., p. 333.

68. Hearings on Military Posture and HR3689, p. 493.

69. If one goes back to the beginning of the armed decoy missile, this event took over seven years to accomplish.

70. Fiscal Year 1977 Authorization for Military Procurement, hearings, Senate Armed Services Committee, Part 2, February 1976, pp. 968-69. A succinct summary of Air Force-Defense Department attitudes toward ALCM is provided in Hearings on Military Posture and HR11500, House Armed Services Committee, Part 5, February-March 1976, pp. 163-73.

71. Fiscal Year 1977 Authorization for Military Procurement, hearings, p. 959.

72. Ibid., pp. 978-79.

73. These numbers and the reasoning behind them were given in Department of Defense Appropriations for 1978, hearings, House Appropriations Committee, Part 2, February 1977, pp. 318-20.

74. The Air Force may have been partly consoled in that it had been given responsibility for the GLCM. One can also note the rather curious managerial arrangement with the Navy in charge of a joint office whose main purpose was to develop an air-launched strategic cruise missile, while the Air Force was responsible for developing a ground-launched version of the Navy's cruise missile.

75. Department of Defense Appropriations for 1978, p. 315. Surprisingly, this reordering of priorities was disputed by Congress. In June 1977 the House-Senate Conference on defense authorizations for FY1978 argued that ALCM-A should take precedence over ALCM-B (A. A. Tinajero, Cruise Missiles [Subsonic]: U.S. Programs, Congressional Research Service, Library of Congress, IB76018, September 1977, p. 13).

76. John B. Walsh, deputy director, Strategic and Space Systems, Defense Research and Engineering, Department of Defense Appropriations for 1978, p. 315.

77. Estimates of the range of the various cruise missiles vary considerably. The figure of 1,250 n.m. for ALCM-A with an external tank is taken from Hearings on Military Posture and HR11500, p. 898. The figure of 1,400 n.m. is based on a comment that the range of ALCM-B would be double that of ALCM-A. Another source, however, states that the stretched ALCM-A would only go 1,100 n.m. and that even this range presumed a high altitude profile for much of the flight (Aviation Week and Space Technology, 4 July 1977, p. 15). Other figures could also be cited. The key point, usually not specified, was whether the estimate of range assumed a high or low altitude flight profile or some mix of the two.

78. About a year later it was stated that the rotary launcher would be designed to permit carriage of both ALCM and SRAM (Department of Defense Authorization for Appropriations for Fiscal Year 1979, hearings, Senate Armed Services Committee, Part 8, February-March 1978, p. 6,809).

79. In 1968, as secretary of the Air Force, Brown had stated: "I continue to believe that an advanced bomber . . . is likely to be desirable at some point" (Hearings on Military Posture and an Act [S. 3293], p. 9,582).

80. In fact, a number of earlier classified studies on bomber penetration had concluded that the standoff carrier would be more effective than a penetrating bomber. The structure and general conclusions of one of these, contracted for by the U.S. Arms Control and Disarmament Agency around 1974, has been described to the author.

The study assumed that the battle would take place over an area approximately 750 miles wide and 1,300 miles deep, that is, roughly one million square miles. One half of the SAM sites was assumed to be concentrated in the first 25 percent of the battle area and the remainder distributed randomly. It was also assumed that the defenses included 400 interceptors with lookdown/shoot-down capabilities. The study then examined the two extremes of (a) full-range bomber penetration and (b) ALCMs launched from beyond coastal defenses, plus various intermediate degrees of bomber penetration and standoff attack. The main conclusions: for all degrees of penetration, cruise missiles prevailed in terms of cost-effectiveness; full-range penetration by bombers produced attrition rates of 25 to 30 percent, uncomfortably high for limited and repeated attack scenarios; and the performance of the cruise missile was critically dependent on its radar cross section. A good AWACS can detect a bomber penetrating at low altitude close to the horizon, which could mean a distance of 300 miles, allowing defense forces a substantial period to attempt an interception. For the U.S. E-3A AWACS, cruising at 30,000 feet, the horizon is about 240 miles away but detection capabilities are believed to extend over the horizon out to about 400 miles (International Defense Review, 1/1977, p. 44). The extent to which the cruise missile would minimize the reaction time available to the defenses depended overwhelmingly on its radar cross section, but also to some extent on its ability to penetrate at lower altitudes than the bomber.

81. Hearings on HR 8390 and Review of the State of U.S. Strategic Forces, House Armed Services Committee, July-November 1977, pp. 89-90.
82. Ibid., p. 89.
83. Ibid., p. 98.
84. Ibid., p. 21.
85. Ibid., p. 83.
86. Ibid., p. 98. The number of B-1s involved is difficult to estimate, but the figure may have been of the order of 118 aircraft since this was the number that top Defense Department officials and members of Congress believed the president would authorize (Aviation Week and Space Technology, 4 July 1977, p. 14).
87. Fiscal Year 1978 Supplementary Military Authorizations, Senate Armed Services Committee, July-August 1977, p. 37.
88. Two weeks after the B-1 decision, Boeing submitted a proposal to the Air Force to use its B-747 in this role. The aircraft was depicted as carrying 70 to 90 cruise missiles, all of which could be launched in 20 minutes (Aviation Week and Space Technology, 5 September 1977, pp. 17-18).
89. This target figure is given in Aviation Week and Space Technology (11 July 1977, p. 15) along with the comment that the coverage of Soviet targets with cruise missiles was more like 70 percent.

90. Aviation Week and Space Technology, 4 July 1977, p. 16.
91. Hearings on HR8390 and Review of the State of U.S. Strategic Forces, p. 101.
92. Aviation Week and Space Technology, 11 July 1977, pp. 15-16.
93. For example, the TU-28P Fiddler interceptor aircraft, first revealed in 1961, has a range of just over 3,000 miles. For several years there have been reports that the TU-28P would be succeeded by an interceptor version of the TU-22 Blinder medium bomber. The latter aircraft is credited with a range of 1,400 miles ("Gallery of Soviet Aerospace Weapons," Air Force Magazine [March 1978]: 94, 98-99).
94. Aviation Week and Space Technology, 11 July 1977, pp. 15-16.
95. Aviation Week and Space Technology, 18 July 1978, pp. 14-15. The Department of Defense is apparently using a standoff distance of 400 to 500 miles for planning purposes (Flight International, 25 November 1978, p. 1,925).
96. This attitude is quite clearly, though indirectly, spelled out in Hearings on HR8390 and Review of the Status of U.S. Strategic Forces, pp. 90-92.
97. International Herald Tribune, 23-24 July 1977, p. 3.
98. Electronic Warfare, Defense Electronics, September-October 1977, p. 55, and Flight International, 12 November 1977, p. 1,411.
99. Aviation Week and Space Technology, 18 July 1977, editorial.
100. Aviation Week and Space Technology, 12 December 1977, p. 13.
101. Aviation Week and Space Technology, 18 July 1977, editorial, p. 7. First reports of a new high acceleration SAM appeared as early as April 1975. At that time it was thought the Soviet Union was designing the new weapon to counter the U.S. SRAM missile (Aviation Week and Space Technology, 14 April 1975, p. 9).
102. Rowland Evans and Robert Novak, "U.S. Faces Cruise Missile Crisis," International Herald Tribune, 31 October 1977, p. 5.
103. Ibid., and Flight International, 12 November 1977, p. 1,411.
104. Aviation Week and Space Technology, 12 December 1977, p. 13.
105. United States Military Posture for FY1978, p. 35.
106. Aviation Week and Space Technology, 20 March 1978, p. 11.
107. International Herald Tribune, 5 June 1978, p. 3.
108. Flight International, 25 November 1978, p. 1,925, and Aviation Week and Space Technology, 20 November 1978, pp. 24-25. In fact Perry had said as much well before the test series was completed. See International Herald Tribune, 5 June 1978, p. 3.

109. <u>Aviation Week and Space Technology</u>, 20 November 1978, pp. 24-25.

110. Harold Brown, secretary of defense, <u>Department of Defense Annual Report for Fiscal Year 1979</u>, p. 119. To the author's knowledge, the last U.S. weapons system that was accorded this status was the Atlas ICBM program in the latter half of the 1950s although, at that time, stating the degree of priority was done systematically and this is not the case now.

111. Ibid. The operative word under the first point is "curb." In August 1977 the distribution of available strategic warheads was SLBMs (47 percent), bombers (30 percent), and ICBMs (23 percent) (<u>Fiscal Year 1978 Supplementary Military Authorization</u>, hearings, p. 9). The ICBM category is the easiest to convert to actual numbers: 2,154 warheads. This implies a grand total of 9,365 warheads of which 4,401 were on SLBMs and 2,809 on bombers, an average of 6.5 weapons per aircraft. The SLBM force is planned to increase by at least 1,920 warheads with 10 new Trident boats (2,400 MIRVs) replacing the 10 boats with Polaris (480 MRVs). Assuming that the 151 B-52Gs are equipped with 20 ALCMs each and the 236 B-52Hs, modernized B-52Ds, and FB-111s continue to be armed with an average of 6.5 weapons, then the stockpile of strategic weapons around 1985 will be 13,029. The SLBM force of 5,321 warheads would account for 49 percent, the bombers for 35 percent, and the ICBMs for 17 percent.

112. Ibid.

113. Ibid., p. 120.

114. <u>Aviation Week and Space Technology</u>, 29 January 1979, p. 22.

115. Ibid., p. 18.

116. Ibid.

117. This intention was also expressed by DDR&E Perry. In reporting the results of the first series of ALCM survivability tests in November 1978, Perry indicated that the proposed wide-bodied cruise missile carrier would only supplement custom-designed strategic bombers (<u>Flight International</u>, 25 November 1978, p. 1,925).

II

EVALUATING THE DEVELOPMENT OF CRUISE MISSILES

INTRODUCTION

Part 1 has essentially described <u>how</u> the strategic cruise missile emerged. It now remains to consider <u>why</u>. Following the distinction made in the Introduction, the discussion to follow will, as far as possible, endeavor to distinguish between why the initial decision was taken to acquire such a weapons system, and why the development program took the particular course it did.

The weapons acquisition process has been studied extensively in the past. From this work it is known that the successful development of a major weapons system is normally the product of the complex interplay of a number of forces. The most obvious of these is a perceived military requirement for the weapon although, particularly with strategic weapons, the "requirement" will often be subjective to some degree. It is convenient to put this force or pressure under the familiar "action-reaction" label, although the U.S.-Soviet military competition has long since progressed beyond the point where a neat pattern of action and reaction can be discerned.

In addition, military requirements can emerge independently of what potential adversaries are up to. This is particularly true in the strategic arena where an enormous and sustained intellectual effort has been made to establish the size, technical capabilities, and employment policy for the nuclear forces that will have the maximum deterrent effect on rival nuclear powers.

Other forces that have been documented in past studies are: technological momentum, which is the desire to develop and produce a weapon simply because it is technically possible to do so; bureaucratic pressure from one or more of the armed services; economic pressures, that is, the need to prevent the disintegration of uniquely skilled research and production teams by providing them with new work (the combination of the last two factors produces the so-called military-industrial complex); and the argument that active weapon development programs are essential as bargaining levers in arms control negotiations.

The following chapters focus on each of these forces with the exception of the two that compose the military-industrial complex. The research for this study revealed no evidence that the "follow-on imperative" was directly significant in the decision to develop strategic cruise missiles. Similarly, it should already be clear from the preceding chapters that sustained and enthusiastic support from the armed services was conspicuous by its absence in the case of strategic cruise missiles.

5
THE MILITARY REQUIREMENT FOR STRATEGIC CRUISE MISSILES

This chapter will be concerned with the military requirements for strategic cruise missiles at two levels. The first is the relatively narrow argument, openly articulated, that the United States needed such a weapon to match the strategic potential of Soviet naval cruise missiles. The second is much broader, namely, the contribution this weapon was expected to make toward alleviating the principal U.S. concerns during the 1970s about the state and trend of strategic balance.

SOVIET STRATEGIC CRUISE MISSILES

The sinking of an Israeli destroyer with SSN-2 missiles and the reemergence of an interest in cruise missiles in the United States led naturally to a great deal more attention being given to Soviet cruise missiles. The existence of the Soviet weapons was used, in part, to justify the U.S. programs, to counter accusations that the United States was starting a "cruise missile race," and to argue that any problems that the planned U.S. cruise missiles posed for arms control already existed.

The continuity of Soviet interest in cruise missiles can be seen from Tables 5.1 and 5.2. The widely accepted view is that the Soviet air-to-surface and surface-to-surface naval cruise missile program stemmed from and was sustained by the threat posed by U.S. naval task forces centered on the aircraft carrier. Throughout the 1950s and probably into the early 1960s, the U.S. Navy had the task of striking targets along the Soviet perimeter with nuclear weapons. Clearly, however, the Soviet arsenal of cruise missiles had a general antiship capability. As the role of the U.S. surface navy in nuclear retaliation declined, as the number of aircraft carriers fell, and as the Soviet Navy began to move farther afield, evaluations of the naval

TABLE 5.1

Soviet Naval Cruise Missiles

Designation	Year Operational	Type of Propulsion	Type of Warhead[a]	Length (feet)	Mission	Platform	Range (n.m.)
Surface-to-surface							
SSN-1 Scrubber	1958	Ramjet	HE	22.5	Antiship	Surface	130
SSN-2a Styx	1959	Rocket	HE	21.5	Antiship	Patrol boats	23
SSN-2b Styx	1964	Rocket	HE	21.5	Antiship		23
SSN-2c Styx[b]	1969	Rocket	HE	21.5	Antiship		23
SSN-3c Shaddock[c]	1962	Turbojet	HE/N	36.5	Antiship	Submarine/surface ship	250
SSN-3a Shaddock	1965	Turbojet	HE/N	38.5	Antiship		(550)[f]
SSN-3b Shaddock	1967	Turbojet	HE/N	33.5	Antiship		
SSN-7[d]	1968	Rocket		23.0	Antiship	Submarine	30
SSN-9	1969	Turbojet	HE/N	29.0	Antiship	Surface ship	40-60 (150)[f]
SSN-11[a]	1973	Rocket	HE		Antiship		23
SSN-12	1976	Turbojet	N	38.5	Antiship	Surface ship/submarine	300
SSN-14	1968	Turbojet	HE/N		Antisubmarine	Surface ship	25
SSN-15[d]	n.a.	Rocket	N		Antisubmarine	Submarine	20
Air-to-surface							
AS-1 Kennel	1956	Turbojet	HE	27.9	Antiship	Badger	55
AS-2 Kipper	1960	Turbojet	HE	32.9	Antiship	Badger	115
AS-3 Kangaroo	1961	Turbojet	HE/N	49.1	Antiship/land attack	Bear	340
AS-4 Kitchen	1965	Rocket	HE/N	37.1	Antiship/land attack	Blinder, Bear, Backfire	150[e]
AS-5 Kelt	1968	Rocket	HE	28.2	Antiship	Badger	120
AS-6 Kingfish	1970	Rocket	HE/N	n.a.	Antiship	Badger/Backfire	150

98

[a] HE = high explosive; N = nuclear.

[b] It is quite probable that the SSN-2c and SSN-11 are the same missile with the former deployed on OSA II-class patrol boats and the latter on modified Kildin and Kashin-class destroyers.

[c] This is some confusion on the number of versions of the SSN-3 in existence and on when they first appeared. It is common practice to put the designation SSN-3a on the version that appeared on converted Whisky-class submarines in 1958 and SSN-3b on the version for surface ships that first appeared in 1962 on Kynda-class light cruisers. The U.S. Defense Department identifies three versions as follows: SSN-3c (1962), SSN-3a (1965), and SSN-3b (1967), and provides the following figures for length: SSN-3c (36.5 feet), SSN-3a (38.5 feet), SSN-3b (33.5 feet). If this information is accepted (as it is in the table), the following sequence seems plausible:

1. The two modes of converted Whisky-class submarines and the Echo I-class, built over the period 1958-62, carried prototype Shaddocks and were not considered fully operational during this period.

2. From 1962 onward, the first definitive version of the Shaddock, the SSN-3c, was deployed on Kynda-class light cruisers, Juliet-class submarines, and Echo II-class submarines. It can also be assumed that the SSN-3c was backfitted to Echo I and Whisky Long Bin-classes of submarine and perhaps even the Whisky Twin-Cylinder class.

3. The longer SSN-3a was probably deployed only on the Kresta I-class cruisers.

4. The latest version, the SSN-3b, being shorter than both earlier versions, was probably backfitted progressively to all modern platforms, that is, the Kynda and Kresta I-class cruisers and Echo II and Juliet-class submarines. One source, however, suggests that the SSN-3b is deployed only on the surface ships (Aviation Week and Space Technology, 2 February 1976, p. 13). It has also been suggested that Juliet and Echo II-class submarines are slated for the SSN-12, the 38.5 feet long successor to the Shaddock (Jane's All the World's Fighting Ships 1977/78). It is worth remembering that the length of a cruise missile can be varied quite easily.

[d] Subsurface launch. The SSN-15 is believed to be the Soviet equivalent of the U.S. Subroc antisubmarine weapon.

[e] One source claims that the latest version of the AS-4 has a range of 425 n.m. when launched from the Backfire, and that this range can be greatly increased if the weapon is programmed to fly a high profile (Aviation Week and Space Technology, 2 February 1976, p. 12).

[f] ( ) indicates considerable uncertainty.

TABLE 5.2

Soviet Naval Surface-to-Surface Cruise Missiles on Major Ships: Platforms (P) and Launchers (L) by Type of Missile, 1965, 1970, and 1978

Class of Ship	1965		1970		1978	
	P	L	P	L	P	L
SSN-1						
Kildin-destroyer	4	8	4	4	1	1
Krupny-destroyer	8	16	4	8	—	—
Subtotal	12	24	8	12	1	1
SSN-3						
Kynda-cruiser	4	32	4	32	4	32
Kresta I-cruiser	—	—	4	16	4	16
Whisky TC-submarine	6	12	6	12	—	—
Whisky LB-submarine	7	28	7	28	7	28
Echo I-submarine	5	30	5	30	—	—
Juliet-submarine	10	40	16	64	16	64
Echo II-submarine	15	120	27	216	27	216
Subtotal	47	262	69	398	58	356
SSN-7						
Charlie-submarine	—	—	5	40	15	120
Papa-submarine	—	—	—	—	1	8
Subtotal	—	—	5	40	16	128
SSN-9						
Nanuchka-corvette	—	—	4	24	20	120
SSN-11						
Kildin-destroyer	—	—	—	—	3	12
Kashin-destroyer	—	—	—	—	6	24
Subtotal	—	—	—	—	9	36
SSN-12						
Kuril	—	—	—	—	1	8
SSN-14						
Krivak-destroyer	—	—	—	—	15	60
Kresta II-cruiser	—	—	3	24	10	80
Kara-cruiser	—	—	—	—	6	48
Subtotal	—	—	3	24	31	188
Total*	59	262	89	498	136	837

*This excludes the missile-armed patrol boats. In 1978 this force numbered 120 carrying a total of 480 launchers for SSN-2b/c Styx. They are excluded because these boats are strictly for coastal defense.

balance moved away from nuclear war scenarios and into the more traditional areas of sea control and the projection of power ashore. In this way, the relatively large number of seagoing platforms with long-range offensive weapons in the Soviet Navy (Table 5.2) began to be viewed as a serious asymmetry.[1] It is important to note that this asymmetry had grown out of a conscious U.S. policy to concentrate offensive naval power on aircraft carriers, not a technological inability to develop antiship missiles.[2]

Thus the decision to develop an antiship missile suitable for widespread deployment on major U.S. surface ships and, subsequently, submarines (the Harpoon) constituted a significant break with past naval policy. Harpoon was designed to be compatible with the launch cannisters used for the Standard SAM and the ASROC antisubmarine weapon, both of which were already widely deployed.

Beyond this general argument, the Soviet weapon that is particularly germane to the U.S. long-range cruise missile program is the SSN-3 Shaddock and its successor, the SSN-12. Military officials in the United States argued that the SSN-3s, at least those deployed on submarines, constituted a potential strategic threat to the United States. Since this perception emerged only after the United States began looking at long-range cruise missiles, it could be viewed as a convenient move to help justify the U.S. programs. In fact, Navy officials argued that it was precisely because the United States was examining cruise missile options that the strategic potential of the Soviet weapons came to be recognized. It was claimed that before this time military officials in the United States, being unfamiliar with cruise missiles, did not fully understand what the intelligence community was saying about the capabilities of Soviet cruise missiles.[3] Thus there was a tendency to equate estimates of the operational range of the SSN-3 against ships with the weapon's maximum range. To a reasonable approximation, this is true for a rocket-powered missile but it is not necessarily true for a cruise missile. The fuel economy of a turbine engine varies significantly with altitude and speed. Moreover, increasing the portion of available volume in a cruise missile devoted to fuel is, relatively speaking, technically uncomplicated.

Once the U.S. Navy had rediscovered the peculiarities of cruise missiles, it was able to appreciate that by changing the flight profile, reducing cruise speed, or by substituting a lighter nuclear warhead for the conventional warhead, the SSN-3 could achieve ranges of strategic significance. Moreover, the outlook could be darkened even further by projecting Soviet technological advances in nuclear warhead miniaturization and in turbine propulsion units. Accordingly, estimates of the range capabilities of the SSN-3 were revised upward. In the late 1960s and early 1970s the most widely quoted figure was 200 to 250 n.m.,[4] while a few years later it was in excess of 450 n.m.[5]

The key point in all this was that one could not tell by external inspection whether an SSN-3 missile was a short-range, antiship version or a longer-range version with a potential strategic role. This point was made with particular force in the case of the new SSN-12. This weapon was credited with a range of 330 n.m. when used as an antiship missile, but was estimated to be able to fly 2,000 n.m. if cruise speed was reduced and if it flew at a higher altitude.[6]

A second key point was that a Soviet cruise missile did not require very long range in order to have strategic utility. The United States is exposed to the open sea on two sides and most of a third side, and its population is quite heavily concentrated in major coastal cities. Thus on the assumption that a cruise missile attack would be launched from 150 n.m. offshore, a 400-n.m.-range cruise missile would put 60 percent of the U.S. population at risk. To pose a similar threat to the Soviet Union, a U.S. cruise missile would require a range of about 1,300 n.m.[7]

A less clear-cut issue was whether or not the Soviet Union assigned a strategic mission to some of its submarine-launched SSN-3s.[8] It is apparently known that some SSN-3s are nuclear-tipped.[9] This would be consistent with an antiship role, particularly with a target the size of an aircraft carrier, but U.S. Navy officials believe that the nuclear warhead on either the SSN-3 or the SSN-12 is too large for any tactical mission.[10] In practice, of course, a demonstration that a particular weapon has (or could be given) some strategic capability is often enough to justify the assumption that it is (or could be) deployed in this role. As it happens, the U.S. Navy's appreciation of the strategic threat from Soviet cruise missiles transitioned in three steps from potential to actual.

In March 1973 the assistant deputy chief of Naval Operation for Submarine Warfare, Admiral Joe Williams, Jr., stated: "we know that these are tactical cruise missiles. But we also are certain that with a modification, which really involves only the warhead of the missile [deleted]."[11] It can be assumed that the deleted portion of this quotation referred to the longer, strategic range the indicated modification would allow.

In April of the following year Admiral Williams's successor, Rear Admiral Synhorst, was saying:

> We should assume that some variants of these missiles have a strategic application. The development of a cruise missile on our part would redress a potential monopoly in these weapons that the Soviet Navy <u>may</u> now enjoy and would provide a stabilizing counterbalance to this Soviet Force.[12]

A year later the SLCM project manager, Captain Locke, simply stated that a strategic version of the SSN-3 existed.[13]

There are several other considerations, however, that diminish the credibility of Soviet submarine-launched cruise missiles as strategic weapons. First, the submarine platforms must come close to the U.S. coast and surface to launch their cruise missiles. The U.S. Navy openly claims an "awesome" advantage over the Soviet Union in the art of detecting and localizing submarines.[14] This is due in part to the fact that Soviet submarines are more noisy than their U.S. counterparts. This would be particularly true of the Echo II and Juliet-class cruise missile submarines because these boats were built 10 to 15 years ago and the missile installations produce notable flaws in the symmetry of the hull. Second, to achieve long ranges both the SSN-3 and the SSN-12 must be programmed to fly at reduced speed and at high altitudes. This, plus the fact that they are relatively large missiles, would significantly reduce their probability of penetration. On the whole it seems unlikely that Soviet planners would be at all confident in the strategic utility of their cruise missiles.

Finally, there is no evidence that Soviet cruise missiles have any great accuracy at long ranges. Although the open literature offers no hard information on this aspect of their performance, the general impression given is that they are decidedly inaccurate. This being the case, the only target for current-generation Soviet strategic cruise missiles (assuming they are assigned a strategic role) would be large cities. One can legitimately question whether a nuclear strike against the opponent's population centers with a weapon that has a relatively low probability of penetration and is launched from a relatively vulnerable platform constitutes a credible scenario, in view of the fact that even an unsuccessful attempt to do this would suffice to justify retaliation in kind. Admiral Holloway, a former CNO, was of a similar opinion. In his words:

> I do not think in a war-fighting sense that [a Soviet attack on cities with cruise missiles] would be a major contribution. It is certainly not the way the United States approaches its nuclear weapons targeting policy.[15]

A more plausible scenario would be a situation in which all the central strategic systems had been expended in a full-scale nuclear exchange, but in which both sides retained their decision-making structures and were engaging in negotiations. In this event nuclear systems with quite marginal strategic capabilities would be pressed into service to provide some leverage in the negotiations.

These various arguments constituted what might be termed the defensive part of the rationale for the U.S. long-range, sea-launched cruise missile. The following section will examine the offensive side of the rationale, that is, the objective U.S. "requirement" for a cruise missile facet to its strategic nuclear forces irrespective of Soviet capabilities in this area. To sum up, while the SLCM was admitted to represent a huge technological advance over the Soviet weapons, the latter were used to argue that the United States was not opening up a new area of strategic competition and that the problems cruise missiles created for distinguishing between nuclear and conventional systems and between tactical and strategic systems already existed, at least in principle.

CRUISE MISSILES AND THE WIDER STRATEGIC BALANCE

The broader question of the contribution the cruise missile was expected to make toward preserving the strategic balance is somewhat more involved. The central observation to be made is that over the period of primary interest to this study—roughly the years 1967-77—the condition of more or less unambiguous Soviet inferiority in strategic nuclear power passed into history. In what follows, the major concerns will be: first, the major developments in the strategic arsenals of the two superpowers and the concerns these developments gave rise to in the United States; and second, the development of thought in the United States on the role of long-range cruise missiles in alleviating these concerns.

One of the more curious episodes of the postwar period was that, despite being the first nation to launch an artificial earth satellite (in October 1957) and thus demonstrating its command of long-range rocket propulsion technology, the Soviet Union did not undertake the large-scale deployment of strategic ballistic missiles until 1966. By this time the United States had almost completed the deployment of 1,054 ICBMs and 656 SLBMs, a program that still ranks as the fastest strategic missile buildup. In fact, as Desmond Ball has shown, these numerical force levels had been decided upon by the end of 1963.[16] An important contributing factor for this delay on the part of the Soviet Union was, ironically, technological inadequacy. It became apparent that the launch vehicle for Sputnik (and also the first Soviet ICBM) was not the result of a breakthrough in rocket technology, but simply an aggregation of many, smaller rocket motors. The first Soviet ICBM, the weapon that lay at the heart of the "missile gap" projections and code-named SS-6 Sapwood, was so huge and unwieldy that only four launchers were eventually deployed.[17]

The first Soviet ICBM to be deployed in quantity was the SS-7 Saddler, which appeared in 1962. This weapon was supplemented, from 1963 onward, by a small number of SS-8 Sasins. In 1966 two new weapons became operational, the SS-11 Sego and the controversial SS-9 Scarp. Over the ensuing five years a total of 1,258 of these missiles were deployed. In addition, between 1968-69 and 1971, 60 units of a fourth ICBM, the SS-13 Savage, were deployed.

In the aggregate, official U.S. estimates of Soviet ICBM force levels went from 340 in October 1966 to 1,520 in November 1971, or from 36 percent to 144 percent of the respective U.S. totals.[18] All these weapons carried single warheads ranging in yield from an estimated 1 MT for the SS-11, to 20 to 25 MT for the SS-9. The Soviet Union's first multiple warhead strategic missile—a version of the SS-11 with three warheads—was deployed in 1973. This was a MRV (multiple reentry vehicles) system and not a MIRV (multiple independently targetable reentry vehicles) system.

In the same year, however, the Soviets began flight testing a new family of four ICBMs, all of which employed the postboost vehicle (PBV) associated with MIRVed payloads.[19] One of these, the SS-18, was a very large missile presumed to be a follow-on to the SS-9. Two others, the SS-17 and the SS-19, were presumed to be competitive prototypes to replace the SS-11, although both were significantly larger than the latter.[20] The last weapon, the SS-16, was a light solid-fueled missile in the SS-13 class. Early in the development program it became apparent that the SS-16 was also deployable as a land-mobile ICBM.[21]

Deployment of these third-generation ICBMs began late in 1974 with the SS-18. The presumption that the SS-17 and SS-19 were competitive prototypes proved to be wrong. Both missiles were deployed from 1975 onward in converted SS-11 silos, although conversion to the SS-19 proceeded much faster than to the SS-17. The status of the SS-16 remained uncertain for some time. Reports that the weapon had gone into production appeared as early as 1976,[22] but two years later it was revealed that the Soviet Union had agreed, in the context of SALT II, to forego production and deployment. It appears the United States argued, and the Soviets conceded, that the SS-16 on its transporter/launcher was indistinguishable from the SS-20, a mobile IRBM not accountable under SALT.

During this period the United States was also active in the ICBM field, although the results of its efforts were notably less conspicuous than those of the Soviet Union. The planned force level of 1,054 ICBMs was reached in 1967 and comprised 54 large Tital IIs, 700 Minuteman Is, and 300 Minuteman IIs. Conversion to the more accurate Minuteman II continued until the Minuteman force was equally divided between the two versions (in 1969). In 1970 the 500 Minuteman Is, the earliest of which was just eight years old, began to be replaced by the Minuteman III, a weapon with three MIRVed warheads, each of which could be delivered with an accuracy three times as great as that of the Minuteman I. By the end of 1974 all the Minuteman Is had been replaced, and during 1975 50 Minuteman IIs were also scrapped in favor of Minuteman IIIs. The Minuteman III itself was also continuously upgraded. Refinements to the NS-20 guidance system progressively lowered CEP below the 1,200 feet estimated for the weapon in its original configuration. In mid-1978 it was decided to proceed with plans to replace the 170 KT-MK12 warheads on 300 Minuteman IIIs with the 350-KT MK12A.[23] These refinements—none of which has any visible impact on the size or shape of the weapon—will significantly increase its lethality.

In the field of SLBMs, the sequence of developments was broadly similar to that for ICBMs. The Soviet Union was the first to deploy submarines with ballistic missiles. This was in 1958 when the first Golf-class, diesel-powered submarine became operational with three SSN-4 missiles. This force eventually totaled 22 units, but the range of the SSN-4 was so limited (about 550 km.) that its strategic significance was at best marginal. The SSN-4 was followed by the 1,300 km.-range SSN-5 Serb in 1963, by which time nine nuclear-powered submarines (the Hotel class) were also available as launch platforms, though still with only three launch tubes per boat. Apart from limited range, a critical weakness of both the SSN-4 and the SSN-5 was that they could not be launched when the submarine was completely submerged.

Thus by mid-1967, when the United States deployed the last of 41 Polaris SSBNs each with 16 tubes for ballistic missiles, the nearest equivalent in the Soviet Union was the 27 SSN-5 launchers on nine Hotel I-class submarines. Like the Soviet Union, the United States had by this time replaced its first SLBM, the Polaris A-1, and was well advanced in converting its force from the second-generation Polaris A-2 to the third-generation Polaris A-3. In the space of four years the Polaris series had progressed from a range of 2,220 km. for the A-1 to 4,630 km. for the A-3. Moreover, the warhead of the A-3 consisted of three MRVs.

Late in 1967 the Soviet Union deployed the first of 34 Yankee-class SSBNs, each with 16 tubes for the 2,400 km.-range SSN-6 Sawfly missile; a longer-range version (3,000 km.) followed in 1973. Some of these longer-range Sawflys carry two MRVs. A new SLBM, the SSN-8 Sawfly-ER (for extended range), also appeared in 1973 deployed on a new 12-tube SSBN, the Delta I-class. The SSN-8 became the longest-range SLBM in the world (over 8,000 km.), although it remained a single-warhead missile. A 16-tube variant of the Delta-class appeared in 1976, and in 1975-76 the first test flights were observed of two new SLBMs, the SSN-X-17 and SSN-X-18, both of which have the PBV associated with MIRVed warheads. Deployment of the SSN-18, which can carry three MIRVed warheads, began late in 1978 on Delta III-class SSBNs.[24]

In contrast to this hectic picture, the United States has deployed only one new SLBM since the Polaris A-3, but this weapon represented an astounding technological advance in its field. The new missile, the

Poseidon C-3, had no more range than the Polaris A-3 but was capable of carrying up to 14 40-KT MIRVed warheads, although most estimates suggest that it is normally deployed with 10 warheads. Between 1970 and 1977, 31 of the 41 U.S. SSBNs were converted to the Poseidon missile, with the remaining 10 boats carrying the Polaris A-3. In addition, the United States is now in the process of building a minimum of 14 Ohio-class SSBNs that will be armed with 24 Trident C-4 SLBMs. It is also planned to retrofit the C-4 into 12 Poseidon SSBNs. The first of these became operational in 1980. The C-4 carries about 8 100-KT MIRVed warheads out to a maximum range of 7,400 km., and is estimated to be 17 percent more accurate than the Poseidon missile.[25]

In the third leg of the strategic forces, the bombers, the Soviet Union has so far not really even tried to match the United States. In the latter half of the 1950s the Soviet Union built approximately 200 Bear and Bison long-range heavy bombers, and from the mid-1960s to the present, official U.S. estimates of the number retained in this role have remained stable at about 140.[26] There is, however, clear evidence that the Soviet Union remains interested in maintaining at least a nominal bomber force in the strategic arsenal. In 1974 a variable-geometry bomber with the code name Backfire was put into production. By the end of 1979 more than 150 were estimated to be operational with production running at 30 to 36 aircraft per year. The performance parameters of the Backfire suggest that it is most suited for a variety of nonstrategic roles, but U.S. officials considered that it had sufficient strategic potential to justify insisting it be limited in some way in a SALT II treaty. Further, since about 1974 rumors have persisted concerning the development of a new long-range strategic bomber; early in 1979 it was revealed that two such aircraft were in advanced development.

In the United States the B-52s remain as the core of the strategic bomber force, although their numbers have fallen from a little over 600 in 1967 to 348 in 1978. In 1969 the 80 B-58 supersonic medium bombers were withdrawn from service and 76 FB-111s procured to replace them. In bomber weaponry, over the period 1972-75, the most significant development was the procurement of 1,500 short-range attack missiles, each with a 170-KT warhead. The next step, which was to have been the B-1, will be the procurement of more than 3,000 ALCMs.

The picture that emerges from this overview is a mixed one, but there is no doubt that in terms of raw strategic power the Soviet Union had brought about a decisive relative improvement in its position by the early 1970s. In numbers of strategic nuclear delivery vehicles (SNDV)—the most simplistic but also the most visible and readily understood index of strategic power—the Soviet Union surpassed the

FIGURE 5.1
Soviet and American Strategic Nuclear Delivery Vehicles, 1967-77

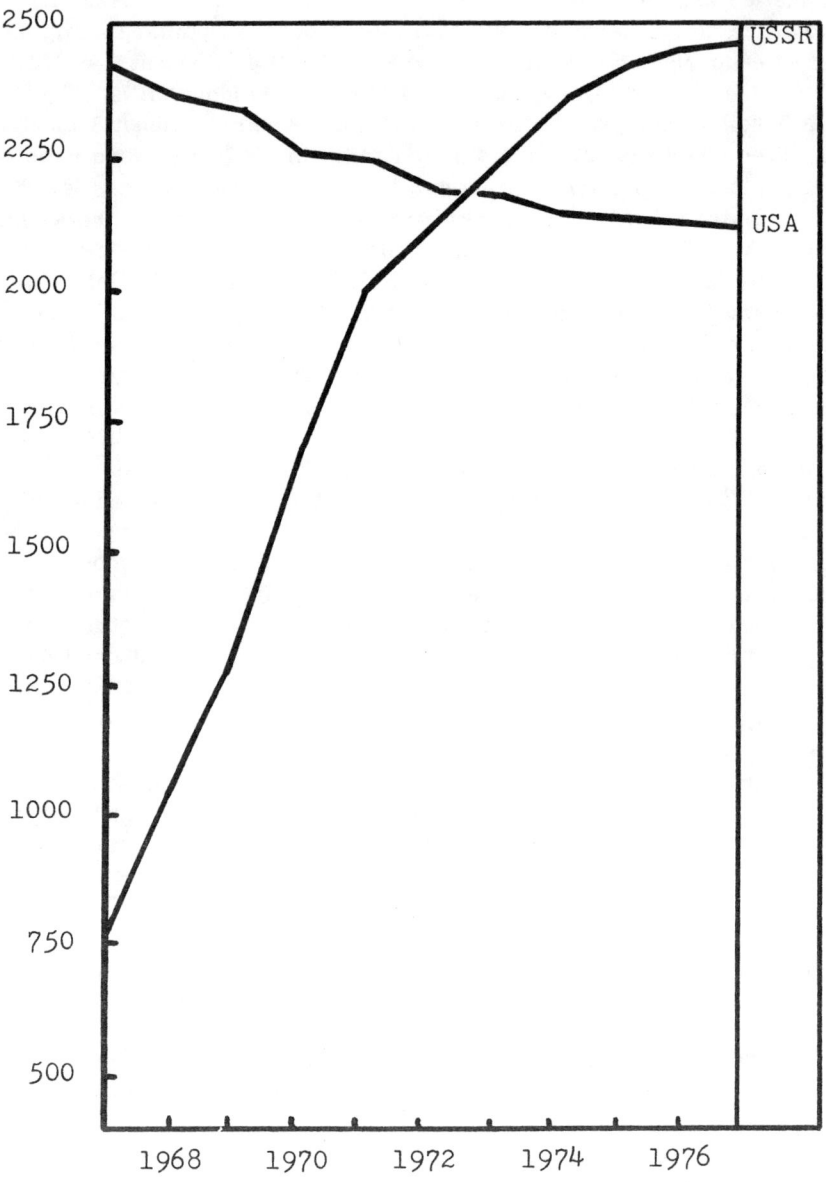

United States during 1972 (Fig. 5.1). In ICBMs alone the crossover occurred in 1970.

WEIGHING THE BALANCE

This unprecedented state of affairs provoked an intense and increasingly divisive debate in the United States in two important areas: on whether the new strategic system in the development pipeline should be accelerated; and whether the attitudes toward nuclear weapons that shaped the performance characteristics of these new developments were still appropriate. Perhaps the most conspicuous product of this debate was the proliferation in the number of indexes used to assess the state and trend of the strategic balance. Most of these indexes have been in use within the military establishment for some time, but their use in the public debate did not become widespread until the early 1970s. This has had positive and negative consequences. The strategic forces of the United States and the Soviet Union are quite unlike in composition. One virtue of a multiplicity of indexes of the state of the balance has been the ability to argue that a condition of "essential equivalence" prevails; that is, one side's superiority in indexes A and B is offset by the other's superiority in C and D. On the other hand, many analysts argue that the various indexes are not equally important. This led to quite contradictory conclusions as to what the United States should do to augment its own arsenal and the sort of constraints on Soviet developments it should seek in SALT.

It is important to stress that many of the numbers that go into these calculations are quite uncertain. The number of SNDVs is probably the least unsure, but even here significant differences arise from: differing estimates on the rate of deployment of new weapons and the rate of retirement of old weapons; and from qualifications such as weapons "operational" on a given date. Definitional problems have also arisen; for example, is the Soviet Backfire bomber an SNDV or, in the case of the B-52/ALCM, is the B-52 or the ALCM the SNDV? On the number of warheads, the advent of MIRV has created an important source of uncertainty, even for U.S. systems. The Poseidon SLBM, for instance, has a maximum payload of 14 MIRVs, but it is widely assumed to be deployed with 10. With a force of 496 missiles this means a potential uncertainty of 1,984 warheads. Similarly, U.S. estimates of the number of MIRVs on the SS-18 were from 5 to 8 in 1975 and are currently 8 to 10. Since payload weight and space can be used for warheads, penetration aids, or devices to improve accuracy, the number of warheads in deployed SS-18s could fall in a wide range. Finally, bombers present a particularly difficult problem in this regard.

FIGURE 5.2
Soviet and American Strategic Nuclear Warheads on Missiles, 1967-77

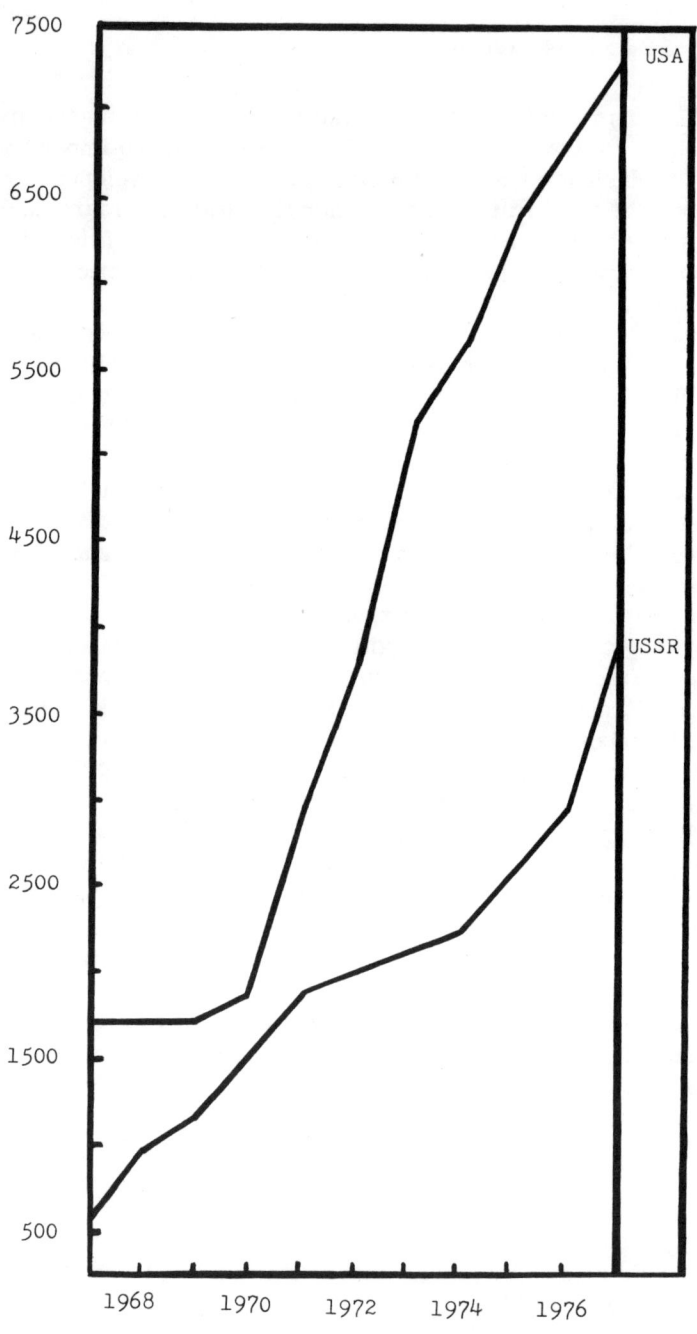

MILITARY REQUIREMENT / 111

Security considerations, plus the fact that since 1963 both countries have tested their weapons underground rather than in the atmosphere, have given rise to notable differences in estimates of the explosive yield of deployed warheads. In the United States the Titan II is credited with a warhead in the 5- to 10-MT range. For the MIRVed version of the SS-19, one source gives 6 × 1-2 MT, another 6 × 200 KT.[27] Other figures for the same weapons could also be cited.

Throw-weight—the total weight of the objects that a missile can deliver to the target area—is an index that has many proponents. It can be estimated with reasonable precision from the dimensions of a missile, although discrepancies still arise. Thus the sources used here estimate 2,500 lbs. for a Minuteman III and 15,000 lbs. for the SS-18, while others estimate 2,200 lbs.[28] and 16,000-20,000 lbs. respectively.[29] As with the number of warheads, bombers present something of a problem here.

Finally, several of the more popular indexes in use are particularly sensitive to the figure used for accuracy or CEP. Tests of long-range missiles can be observed, but since the observer does not know the exact aim-point, his estimate of the weapon's accuracy is inherently uncertain. In addition, the point is sometimes made that since the United States conducts relatively few missile tests and never from operational silos, its confidence in the operational accuracy of its own weapons is relatively low. The Soviet Union, because it has conducted more than 100 tests per year in recent years, including many from operational silos, will be less uncertain on the operational accuracy of its missiles.

With these preambular remarks, one can now turn to the various indexes and how the two countries measure up in relation to them. An important performance parameter is the number of separate targets that each side can hit and, if the targets are "soft," destroy. The obvious index here is the number of individually targetable warheads. Leaving aside the bombers, Fig. 5.2 shows that diminution of the U.S. lead was abruptly arrested in 1970 as the MIRV program got under way. The gap widened through 1974 and has since shrunk, primarily because the Soviet Union began to deploy MIRVed missiles in 1975. However, as of the end of 1977 the United States retained a commanding lead—some 10,000 versus 4,100 if the estimate for bomber weapons in 1977 made above is added. The limitation of this index derives from the fact that many, if not most, of the targets of the strategic forces are to varying degrees "hard," even in relation to a nuclear blast. Thus a distinction must be drawn between the number of targets attacked and the number destroyed, with the latter critically dependent on accuracy, the degree of hardness of the target, and the yield of the warhead. These additional variables go into some of the more complex indexes, which will be considered in a moment.

FIGURE 5.3
U.S. and USSR ICBM and SLBM Forces:
Equivalent Megatonnage, 1967-77

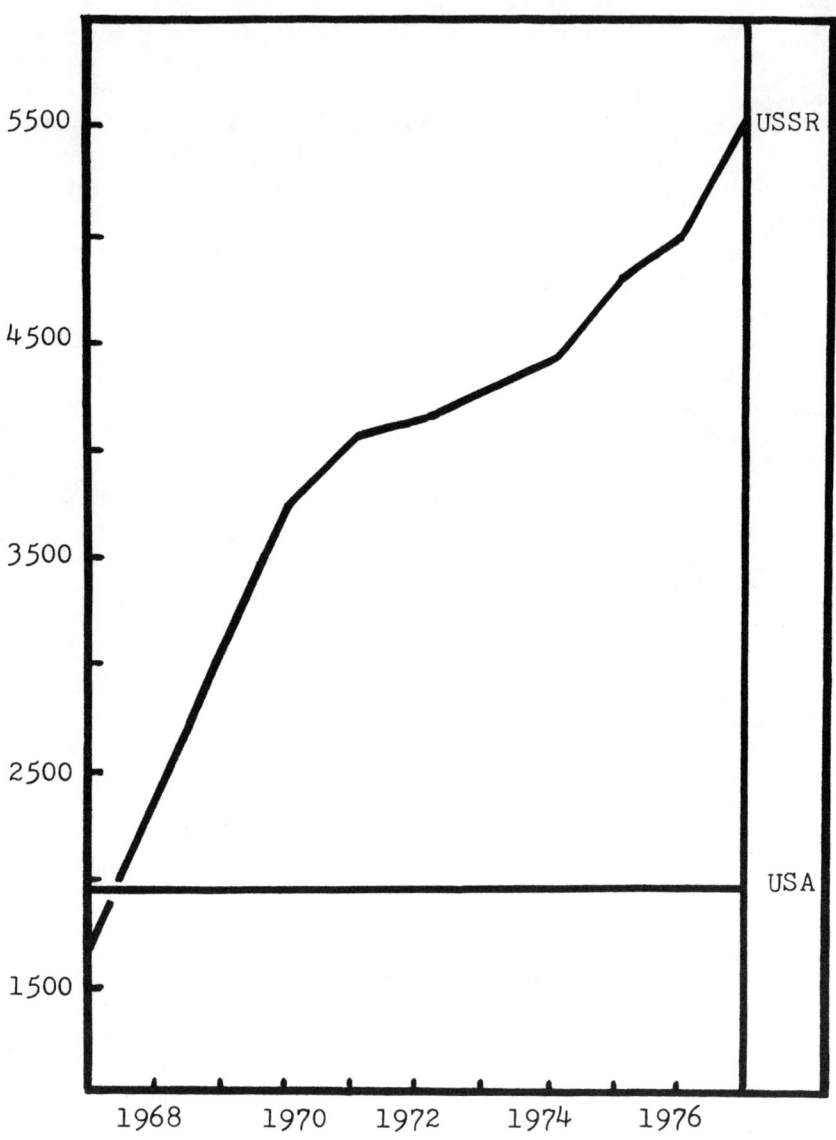

Another relatively straightforward index is the total megatonnage loaded aboard deployed SNDVs. If nuclear weapons are fused so that the fireball reaches the earth's surface, their aggregate yield bears a direct relation to the amount of radioactive fallout produced. If the purpose of nuclear weapons is to threaten the opponent's population, then the amount of fallout created is a relevant measure of strategic power. On the other hand, if the purpose is to destroy manmade structures, total megatonnage is a misleading index. This is because the area exposed to a given level of damage from the blast of an explosion does not increase in direct proportion to the yield of a weapon, but roughly in proportion to the two-thirds power of its yield.

Adjusting for this fact produces a measure called equivalent megatons (EMT). Bearing in mind that the precise yields of nuclear weapons are not known, particularly Soviet ones, Fig. 5.3 compares the EMT of the ICBM and SLBM forces for the two countries. For these forces the Soviet Union moved ahead on this index in 1968. If bombers are added the Soviet lead would be reduced but not eliminated. According to the U.S. Defense Department, when all strategic forces are included, the Soviet Union has been ahead on EMT since 1970.[30]

In itself EMT has obvious limitations. First, it gives no indication of the number of separate targets the respective forces can attack. Second, most high priority targets are hardened and the destruction of such targets is far more sensitive to the accuracy of a weapon than to its yield.

The final single-parameter index is throw-weight or payload. As pointed out earlier, the throw-weight of a missile can be estimated with reasonable precision, although bombers present a more difficult problem. On the basis of the author's data, the Soviet Union moved ahead on total throw-weight in 1972 (Fig. 5.4); the U.S. Defense Department shows this occurring in 1975 or 1976.[31] A partial indication of the general accuracy of the data is that, in mid-1977, total U.S. throw-weight was officially stated to be 75 percent that of the USSR.[32] The author's figures put the relationship at 78 percent in 1977.[33]

The significance of a throw-weight gap depends a great deal on relative technical capabilities. The side with the more advanced technology can do more with a given payload, whether in terms of EMT, number of warheads, or the accuracy of individual reentry vehicles. At the present time, the United States clearly has technological leadership in the relevant areas, for example, light compact warheads and the miniaturization of electronic components for guidance systems. However, many observers are concerned that as its technology improves, the Soviet Union will have more payload available to take advantage of these improvements. The background to this concern is that the preoccupation with numbers in the SALT negotiations is likely to leave the United States permanently inferior in terms of throw-

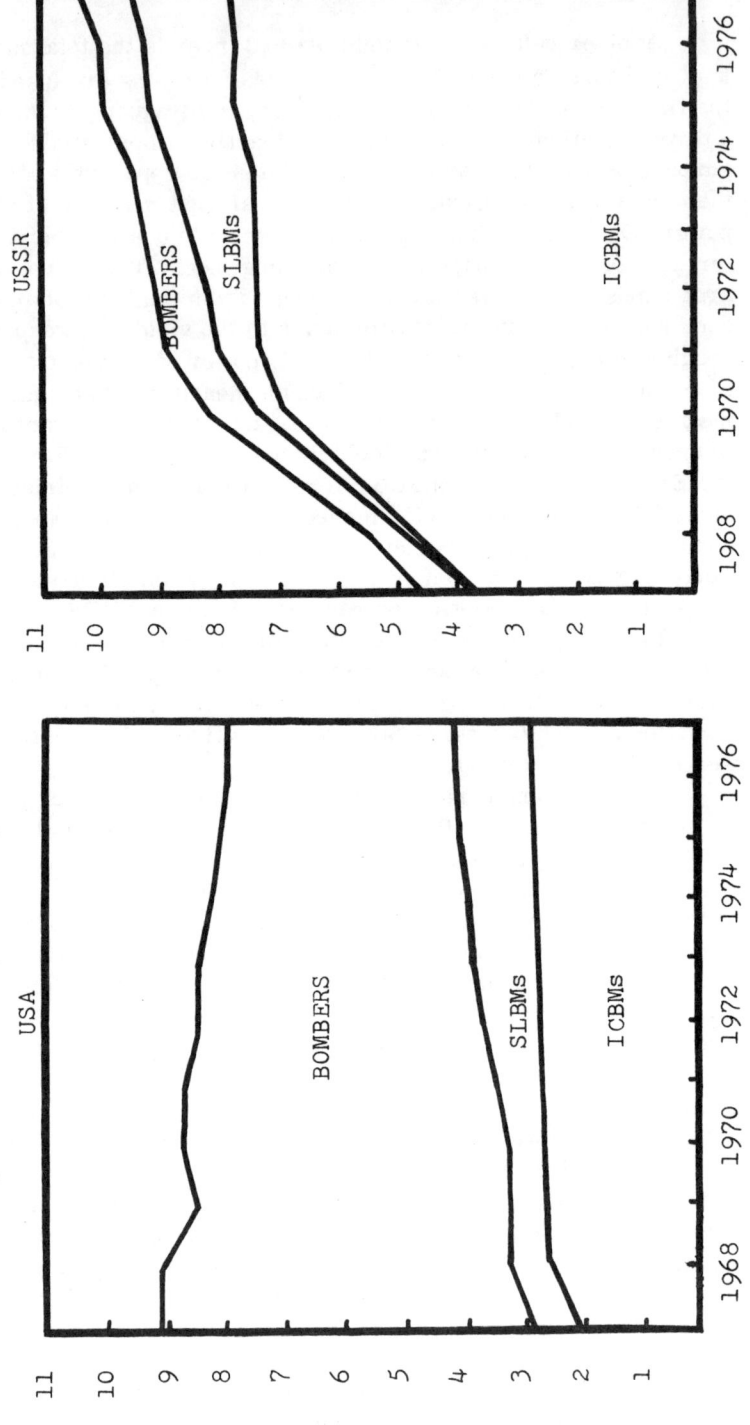

FIGURE 5.4
Throw-Weight of Soviet and U.S. Strategic Nuclear Forces, 1967-77 (millions of pounds)

MILITARY REQUIREMENT / 115

weight, while Soviet technological improvements will be unchecked.
A related concern, highlighted in Fig. 5.4, is that the bulk of Soviet payload resides in ICBMs (75 percent in 1977). ICBMs are fast, accurate, can readily be maintained at high rates of alert, and can be kept under secure command and control. They are therefore regarded as the best first-strike weapons.

The key deficiency of all these single-parameter indexes is that they give no indication of the ability of the respective forces to destroy hard targets. An indicator specially designed to measure this ability—variously called lethality, countermilitary potential, or hard-target kill potential—takes the EMT of each independently targetable warhead and divides this by the square of its accuracy or CEP. The resulting numbers can be added.

All computations of lethality in the open literature show the United States leading by a substantial margin.[34] At the same time the respective analysts differ quite markedly in their estimates. This is mainly because lethality is highly sensitive to accuracy, which is perhaps the most difficult parameter of missile performance to determine. The three analysts just cited, all writing in the period 1974-75, varied in their estimates of the lethality of the Soviet ICBM forces by a factor of nine, with the differences due almost entirely to the CEP estimates used in their calculations. It is worth noting again that differing opinions on the accuracy of various weapons are not simply the result of the fact that, for security reasons, the correct figures are not published. For reasons cited earlier the classified figures are also unreliable, so that even an index of lethality computed for officials with the highest security clearance would have to be treated with circumspection.

This is an important point because some analysts refine the lethality index so that the available lethality can be compared with the lethality required to destroy the enemy target system with a high probability of success. These calculations require a new input, namely, the degree of hardness of the various targets. Once again neither side can be fully confident in its estimates of the extent to which the opponent's targets are hardened. These various uncertainties are compounded by an effect known as fratricide. To raise the statistical probability of successfully destroying a hard target to a high level, it will often be necessary to send more than one warhead against the target. Most observers seem convinced that the explosion of the first weapon will create so much dust, debris, radiation, and turbulence that follow-on weapons will be thrown off target and may not even detonate.[35]

The lethality index can be misleading if, say, the ICBM force is composed of various types of missiles. A very high number for lethality may be concentrated in a relatively small number of high performance missiles, but no matter how good they are these missiles can only attack a finite number of targets. For example, if the CEP of the 288 SS-9 ICBMs that the Soviet Union deployed through 1974 is reduced from 1 n.m. to 0.3 n.m., the lethality of this force (assuming a single 20-MT warhead) increases from 2,122 to 23,578, but it can still strike only 288 targets. As a final precautionary remark, it must be pointed out that the computations of lethality that appear in the open literature are all confined to missile forces, ICBMs and SLBMs. Moreover, computations of the lethality required to destroy the opponent's forces are restricted only to ICBMs. This arises because missile lethality and the lethality required to destroy fixed ICBM silos can be precisely computed (though many uncertainties are hidden in the numbers). Bomber lethality presents problems because there are so many variations on possible weapon loadings and because bombers face active defenses. Similarly, calculations on how many bombers might be caught on the ground and how many SSBNs in port depend on variables that are difficult to quantify.

In short, as an indicator of strategic capability, lethality has both advantages and shortcomings, as do all the other indexes. It has, however, won official acceptance as a measure of the strategic balance. In mid-1977 Defense Secretary Brown stated that the United States led the Soviet Union by a factor of 1.6 in hard-target kill potential or lethality.[36] This estimate covered all strategic forces—ICBMs, SLBMs, and bombers—and since the United States has a huge lead in bombers it seems that the Soviet Union is rapidly closing the gap on this index. This follows from the fact that, in 1974, Tsipis put the United States ahead in missile lethality by a factor of 5.9; even Congressman Leggett, who assumed greater accuracy for most Soviet missiles, concluded that the United States led by a factor of 3.3. In view of the controversy in the United States over the prospect of a Soviet lead in counterforce capability sometime in the 1980s, it is worth stressing that, by that time, the Soviet Union will have lived with the reverse situation for at least 20 years. Moreover, for all those 20 years the bulk of Soviet strategic power resided in fixed land-based ICBMs, the weapons most vulnerable to a counterforce strike. Even now ICBMs account for less than one quarter of the warheads in the U.S. strategic arsenal, and this fraction is programmed to fall over the next decade.

A final indicator that has recently gained some currency is equivalent weapons (EW).[37] The targets for strategic weapons probably fall into three categories—hard point-targets, soft point-targets, and soft area-targets—and for each category a particular index is most appropriate to measure force effectiveness. The EW index combines the most appropriate indexes for each of the target categories into a single measure of force effectiveness. While this approach has obvious virtues, its use by analysts without access to classified information will almost certainly produce results even more varied than those just noted for lethality. This is because EW requires that the analyst also must estimate the target plan each side has prepared—the SIOP in the case of the United States—and this information is highly classified. Moreover, the target plan will be highly sensitive to the scenario assumed: who strikes first, the scale of the attack, and so on.

DEBATING THE BALANCE

The scorecard reads as follows: United States ahead in the number of warheads and hard-target kill capability, but with its lead in both areas diminishing since about 1975; the Soviet Union leads in delivery vehicles, throw-weight, and equivalent megatonnage. In the latter two areas the Soviet lead was still growing in 1977. In consequence, those who argued that the Soviet Union appeared to be seeking strategic superiority and would achieve it unless the United States responded, found an increasingly receptive audience. Conversely, those who rationalized Soviet strategic developments as a response to an unnecessarily large U.S. buildup in the early 1960s, coupled with a technology that could only produce large missiles with large warheads, faced an increasingly skeptical audience.

The extraordinary scope and continuity of the Soviet strategic program made more people at least curious as to what its purpose might be. It also made them concerned as to whether the United States was doing enough to preserve the substance and appearance of essential equality in strategic capability. Opinions in the United States also appear to have been influenced strongly by the feeling that the Soviet Union had moved to extract the maximum possible capabilities out of the letter of the SALT I agreement and, in the process, had shown little sympathy for the spirit of that agreement. For example, it is replacing its SS-11 ICBM—which the United States understood to represent the outside limit of a "light" ICBM—with the SS-19, which is 60 percent larger in volume and can deliver a payload 350 percent heavier than its predecessor. Other instances included delays in retiring old ICBMs as they were replaced by new SLBMs, and the apparent deployment for a period in 1977 of more than the 62 "modern" SSBNs allowed under SALT I. In every case the issue was resolved, but the impression given of operating within the law, but only just, undoubtedly contributed to the growing inclination to view Soviet intentions with more than the usual suspicion.[38]

These concerns, although expressed most forcefully by observers outside government, were shared by many leading officials. In his annual report for FY1979, Defense Secretary Brown professed himself to be uncertain as to why the Soviet Union was "pushing so hard" to improve its strategic nuclear capabilities. He further stated that the United States could not afford to assume that the Soviet Union was "motivated by considerations . . . of pure deterrence."[39] At least in part, the latter comment was probably directed at the increasing vulnerability of the U.S. ICBM force to a Soviet strike. To quote Brown: "A relatively small fraction of the current generation Soviet MIRVed ICBMs could, by the early-to-mid-1980s, reduce the number of surviving Minuteman to low levels."[40] Brown stressed the enormous uncertainties the Soviets would have to contend with in planning and executing such an attack and in forecasting the U.S. response, but he also acknowledged the "political cost" involved in allowing the Soviet Union to gain or appear to gain an unmatched capability of this kind.[41]

This assessment was a middle-of-the-road position, acknowledging the validity of arguments on either side. Colin Gray, for example, has argued:

> On the Soviet side, the emergence over the next decade of a total hard target counterforce capability against upgraded Minuteman silos is not seriously in question. What is in question is precisely when the Soviet Union will approach such a level of competence.[42]

At the other extreme are those who challenge whether a total countercapability against fixed ICBM silos will ever emerge because of the fratricide problem. Still others acknowledge the theoretical possibility of a successful first strike against ICBMs but, like Secretary Brown, stress the enormous uncertainties involved and point out that "categorical assumptions about vulnerability . . . rest upon tacit assumptions more than technical fact; and, that the usual assumptions are not the only ones which ought to be made."[43]

Thus most analysts were preeminently concerned with the state of the strategic balance in the future. Few suggested that the situation, even in the latter half of the 1970s, represented anything other than essential equivalence with a healthy margin of crisis stability, that is, neither side could significantly improve its relative position by striking first at the other's strategic forces. The problem, it was contended, resided in the fact that the momentum of Soviet strategic developments far exceeded that of the United States. During the 1970s the United States had dramatically increased the effectiveness of a quantitatively static strategic force by pursuing qualitative improvement, primarily MIRV and higher accuracies. The Soviet Union, in contrast, had concentrated on assembling a force that, in quantitative terms, was significantly larger than that of the United States, and therefore had more "space" available for force-multiplying qualitative improvements.

Accordingly, interest focused on the relative capabilities of the strategic missile forces the two sides could deploy in the 1980s under the SALT constraints then foreseeable. In this context, the key issue was the relative vulnerability of the two countries to a first-strike counterforce attack. This particular issue had been gathering force since the appearance of the SS-9 in 1966; by the mid-1970s, with the deployment by the Soviet Union of its MIRV generation of ICBMs, it virtually monopolized the strategic debate in the United States.

In counterforce scenarios the key weapon is the ICBM. As has been seen, the Soviet strategic forces are utterly dominated by the weapon. Moreover, there was little reason to expect any major change in this distribution over the period of a SALT II treaty, that is, through the end of 1985. Given that the U.S. ICBM force was expected to remain unchanged over this period in terms of throw-weight and numbers of missiles and warheads, this meant the Soviet lead in ICBM warheads and throw-weight could be expected to increase almost continuously.

Until about September 1977, force projections were predicated on the assumption that a SALT II treaty would adhere closely to the numbers agreed to at Vladivostok, namely, a ceiling of 2,400 on strategic delivery vehicles and a subceiling of 1,320 on MIRVed missiles. In the mid-1970s, it was not unreasonable to assume that the Soviet Union would deploy 1,320 MIRVed ICBMs—SS-17, 18, and 19. It could be argued that this implied a MIRVed ICBM force with a total throw-weight of some 11 million lbs. capable of delivering at least 7,500 1- to 2-MT warheads.[44] The opposing U.S. ICBM force would, in the mid-1980s, still consist of 1,054 launchers, with a throw-weight of 2.9 million lbs distributed over 2,154 independent warheads. The revised SALT II ceilings and sublimits tentatively agreed upon in September 1977 did improve the picture somewhat from the U.S. perspective, but a large Soviet lead in the early 1980s in prompt counterforce capability remained virtually certain. Specifically, with the new sublimit of 820 on MIRVed ICBMs, the Soviet Union could still be expected to deploy well in excess of 5,000 warheads on these weapons alone.

In the apolitical calculus of counterforce, the significance of this imbalance can be portrayed as follows:

> Until Soviet accuracy is improved to better than 0.2 mi., it will be difficult for them to eliminate more than half of our Minuteman silos on an initial strike, even if they target two of their RVs [reentry vehicles] on each silo. When their accuracy approximates 0.15 of a mile, around 90 percent of U.S. silos would be vulnerable to such a two-on-one attack. A two-on-one attack would require less than half of the MIRVed ICBM RVs they are expected to have available by 1985. If and when their accuracy approx-

imates 0.1 mi., around 90 per cent of our silos would be vulnerable to an attack by a single RV against each silo, provided additional RVs are programmed to substitute for missiles that fail during their launch phase. If we use all of our Minuteman IIIs—assuming improved accuracy and the substitution of the MK. 12A for the MK. 12 RV— against Soviet silos, it is unlikely we could destroy more than 60 per cent of them.[45]

Reactions to this emerging state of affairs varied. Those inclined to discount the possibility that either side really would assume the huge risks associated with a first strike still argued that it was vitally important that all concerned perceive the strategic forces to be in balance. On this score the magnitude of the emerging imbalance in ICBM forces was unacceptable. It is appropriate to stress the word "magnitude" because the Soviet Union had surpassed the United States in ICBM warheads in 1976, in launchers in 1970, and in throw-weight in 1966 or 1967.

Others argued that this situation had come about because the United States had been unduly preoccupied with the number of launchers or delivery vehicles throughout the SALT exercise. It now urgently needed to switch its emphasis to throw-weight or numbers of warheads. A third reaction was more disturbing. It was argued that by the early to mid-1980s, the U.S. ICBM force would no longer deter an attack. It was further argued that the Soviet authorities might calculate that the exchange ratio (RVs expended versus RVs destroyed) in a first strike on the ICBMs, bomber bases, and SSBN ports was sufficiently favorable to produce a postattack balance of forces so heavily weighted in their favor that a U.S. president would be deterred from authorizing any kind of retaliation.

Perhaps the most forceful exponent of the view that the viability of the U.S. strategic deterrent would become increasingly open to question was Paul Nitze. In a short article in Foreign Policy, Nitze argued that the real test of stability of the nuclear balance was not a comparison of the forces in being at any given moment, but whether a balance still prevailed after a counterforce exchange (that is, after each side had destroyed as many of the other's offensive nuclear weapons as possible while keeping in reserve adequate weaponry to threaten a counterpopulation strike).[46] Nitze presented a number of graphs on which were prescribed the state of the strategic balance in terms of a number of indexes—throw-weight, EMT, warheads, and so on—over varying periods of time. Three situations were compared: (a) prewar, (b) after a Soviet counterforce strike, and (c) after a U.S. counterforce response. Nitze showed that at stages (b) and (c) the Soviet Union would increase its relative superiority, and after 1980 a counterforce exchange would leave that country with a substantial superiority in every index of strategic power. Most particularly, the Soviet Union would be farthest ahead on remaining throw-weight, which is the best measure of countervalue potential. Remaining U.S. forces, because of the long-standing preference for smaller and more accurate weapons, would be small in terms of throw-weight and megatonnage. The casualties these could inflict on a dispersed and sheltered Soviet population would be relatively low and quite possibly acceptable to the Soviet leadership.

Nitze did not specify in any detail the assumptions on which his calculations were based, so these cannot be challenged.[47] The various considerations mentioned earlier that militate against the practical feasibility and likelihood of a first strike can be invoked to temper the severity of his conclusions. His assumption that U.S. forces would be caught on day-to-day alert can be challenged. The same goes for his assessment of the effectiveness of the Soviet civil defense system.[48] The point, however, is that Nitze's views were shared by an influential segment of the strategic community in the United States and, to some degree at least, probably by everyone. That is, the aspect of the strategic nuclear balance that was of preeminent concern was the growing vulnerability of silo-based ICBMs and the prospect that the Soviet Union might soon have a considerably greater counterforce potential than the United States.

By the beginning of 1978, if not earlier, there was no longer any dissent from the view that the survivability of the Minuteman force would be reduced to low levels in the early to mid-1980s. Secretary of Defense Brown said so without qualification. He also admitted that a SALT II agreement would not prevent this development.[49] To an important extent, the significance attributed to this development hinged on how intimately individual analysts linked the strategic nuclear forces with the pursuit of foreign policy objectives or, in other words, how one answers Henry Kissinger's now-famous rhetorical question, "What in the name of God is strategic superiority?" At one extreme are those who view the link as a weak one, or at least advocate that the link should be weak. These analysts are prepared to accept a quite marked asymmetry in strategic forces, provided only that U.S. forces retain the ability under all conditions to inflict massive damage on Soviet population and industry. At the other extreme are those who perceive the link to be unavoidably an intimate one. Strategic forces may be the weapons of last resort, but they are that and one can construct plausible (or not implausible) scenarios in which a credible threat to employ them or even their actual (limited) use will be necessary in order to avoid defeat on vital issues. For these analysts the probable nature of the strategic balance in the mid-1980s was profoundly disturbing because at least on the basis of their own computations, the United States could not escalate out of an impending military disaster (say in Europe) reasonably confident that the Soviet Union would be unable or unwilling to match or overmatch the U.S. escalation.[50]

One clear indicator of the widespread support for Nitze's views was that, in his first annual report, Defense Secretary Brown opted to present his own assessment of how the strategic forces would shape up after a counterforce exchange.[51] The index used by Brown was a nonspecific "relative force size," a measure of the ability to destroy a given set of economic and military targets. Like Nitze, Brown showed the situation prevailing before an attack, after a Soviet first strike, and after a U.S. counterforce retaliation, with both sides on normal day-to-day alert. Relative to Nitze's indexes, Brown's assessment of relative force size was far more favorable to the United States in every situation, but especially that following a counterforce exchange. Moreover, Brown argued that if the calculations were done assuming both sides to be on generated alert, the outcome would be even more favorable to the United States.

Thus there are two conflicting assessments of the outcome of a nuclear counterforce exchange at any time between the mid-1970s and the mid-1980s. One is absolutely official but employs a rather vague index of strategic strength, and the other is based on classified, and therefore presumably officially supplied, data. To a very important extent, these conflicting assessments reflect the flexibility of the rules under which the numbers game can be played, particularly the counterforce variant of this game. It probably has not escaped the reader's notice that in the preceding discussion no reference was made to the counterforce potential of SLBMs. It is generally assumed that inherent limitations in precisely determining the position of the submarine at the time of launch will prevent SLBMs from achieving any worthwhile counterforce potential until the ability becomes available to maneuver the warhead onto the target with some form of terminal guidance. The development of maneuverable reentry vehicles (MARV) has been under way for several years; but the conventional wisdom that until something like MARV is perfected SLBMs will be strictly countervalue weapons is not indisputable.[52] In the areas that have the most direct bearing on the counterforce potential of SLBMs—the accuracy of the missile and of the navigation equipment on the launching submarine, and the speed and reliability of communications with the submarine—the research and development effort has been intense and continuous.[53] Moreover, the U.S. Navy has long had the goal of giving its SLBMs a counterforce potential, as the following statement by Secretary of the Navy Tyler Marcy makes clear:

> For us it seems to me that for the United States through

time to be able to upgrade its seaborne forces, so that they have hard target capability in an analogous way to those that we spent [sic] for our land based strategic forces, would be vitally important to us as a nation.[54]

If even current-generation U.S. SLBMs have a worthwhile counterforce potential (and if anyone believes this it will be the Soviet defense planners), the impact on the comparative lethality of the opposing strategic forces would be quite dramatic. Currently the United States has nearly six times as many SLBM warheads as the Soviet Union and there is little prospect—in view of the Trident program—that the magnitude of this lead will be significantly reduced over the next decade.

Thus one possible explanation of Secretary Brown's more comforting (compared to Nitze's) assessment of the counterforce potential of U.S. and Soviet strategic forces is that: his inputs regarding accuracy, reliability, target hardness, and so forth were relatively more favorable to the United States; and he might have given some credit to the (growing) counterforce potential of U.S. SLBMs.

A second line of explanation would recognize that, in the 18 months or so that separate these assessments, two influential developments took place:

Some significant changes and additions were made to the counting rules for SALT II compared with the Vladivostok guidelines; and

The cancellation of the B-1 in favor of the B-52/ALCM combination plus the possibility, if the need arose, of a fleet of wide-bodied cruise missile carriers.

The Carter administration had made it clear that it was staking a great deal on the B-52/ALCM combination to preserve a strategic balance during the term of SALT II. Its officials argued strenuously that the weapon would be equal to the task. In testimony to the House Armed Services Committee in August 1977, Secretary of Defense Brown presented the following projections.[55]

(a) A B-52/ALCM force would preserve U.S. superiority in number of warheads through 1986 although the margin would shrink from the prevailing 2.4 to 1.26 that of the Soviet Union. If a fleet of wide-bodied cruise missile carriers (CMC) were added U.S. superiority would drop only to 1.87;
(b) U.S. throw-weight, currently 75 percent that of the Soviet Union, would drop to 48 percent in 1986 even with the B-52/ALCM force but could be restored to

77 percent if a fleet of CMC were added before that date;

(c) In hard target kill potential the United States had a prevailing superiority of 1.6 times that of the Soviet Union. The B-52/ALCM force alone would not prevent the emergence of a Soviet superiority of 1.5 times that of the United States by 1986. However, the addition of a CMC fleet would restore a U.S. lead of 1.68 times that of the Soviet Union.[56]

An independent depiction of the impact of ALCMs on the stability of the strategic balance is shown in Fig. 5.5. The line for neutral stability describes the situation in which neither side could emerge from a counterforce exchange with a significant preponderance of strategic forces. The very high accuracy of the cruise missile would help enable the United States to destroy enough of the Soviet forces held in reserve to reinstitute a balance after a Soviet first strike. The diagram is based on a force of 2,500 cruise missiles, that is, the B-52/ALCM force alone.[57]

It appears to have been recognized quite early in the cruise missile development program that confidence in the ability of these

FIGURE 5.5
Influence of ACLM on Strategic Stability

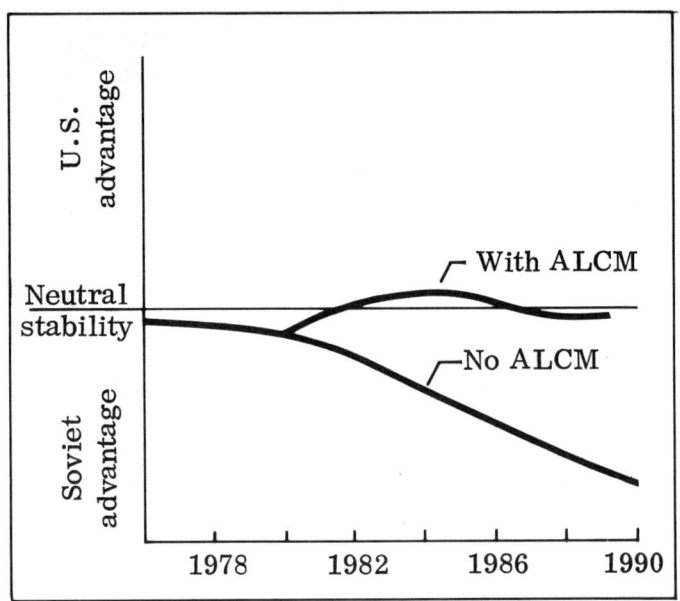

Source: Peter Hughes, "Arms Control and Strategic Stability," Air Force Magazine (April 1978): 61. ©1978 by the Air Force Association.

weapons to penetrate to their targets rested as much on the magnitude of the attack as on the capabilities of the weapon itself. The cruise missile—with a tiny radar cross section, low infrared signature, and penetrating at very low altitudes—is beyond doubt a very difficult weapon to counter. But it is not an insurmountable difficulty: it is still a weapon that must fly for two hours or more at subsonic speeds over, under, or around enemy air defenses. This fact must dramatically reduce confidence in the ability of any one weapon to penetrate. High confidence results if the good penetration characteristics of individual missiles are combined with what the Air Force calls high "mass flow," that is, large numbers of missiles penetrating simultaneously. Given the available strategic weapon platforms in the U.S. arsenal, the quickest and least expensive way of threatening the Soviet Union with a massive cruise missile attack was to modify a large number of B-52s to carry 20 missiles each.

Secretary Brown's confidence in the effectiveness of the B-52/ALCM force and the additional leverage provided by the CMC option can be, and of course has been, challenged on several grounds. If the stability of the strategic balance—in the sense of essentially equal forces after a counterforce exchange—is left heavily dependent on the B-52/ALCM force, then future developments in Soviet air defenses become a major potential source of instability. This line of argument was not particularly troublesome to Carter administration officials. As has been seen, it was readily admitted that the Soviet Union could, in the foreseeable future, deploy air defense capabilities that would exact high rates of attrition on a cruise missile attack. The major components of such a defense system are considered to be: (a) a force of 50 to 100 AWACS aircraft coupled with up to 3,000 high performance interceptors with lookdown/shoot-down capabilities; and (b) deployment of 500 to 1,000 SA-10 SAM sites with associated radars and command and control complexes. Component (a) would concentrate on destroying cruise missiles en route to their targets, but might also endeavor to attack B-52s offshore before they could launch their missiles; (b) would constitute the terminal defenses around high-value targets. On the other hand, it is estimated that these forces would cost $30 to $50 billion and could not be fully deployed before about 1985 at the earliest. Moreover, if the Soviets acquire component (a), the United States will have the ASALM weapon available to attack Soviet AWACS aircraft. More generally, the United States can improve the penetration capability of its cruise missiles in a number of ways; for example, make them faster or add ECM and a reactive maneuvering capability that would function automatically if the missile sensed it had been detected by Soviet radars.

A more telling line of criticism or concern was that, if the administration was to an important extent basing the substance and appearance of an adequate strategic deterrent in the 1980s on the B-52/ALCM, then the concessions made in the negotiations on a SALT II treaty were, to say the least, unwise. Among the details of the prospective SALT II treaty released in September 1977 was a U.S. agreement to limit ALCM range to 2,500 km. and to count B-52s equipped with ALCM as MIRVed delivery vehicles. It is true that the range limitation would have gone into the three-year protocol attached to the treaty (it was eventually agreed that no restrictions would be placed on ALCM range), but it was equally true that the United States would have been under considerable pressure at the end of three years to retain that restriction. A permanent restriction on ALCM range would make the erection of a forward air defense barrier more attractive for the Soviet Union because the payoff would be reduced target coverage for the cruise missiles launched beyond the barrier. Similarly, counting B-52/ALCMs (and CMCs) as MIRVed systems amounted to a significant additional restraint on increasing the number of cruise missile platforms since high-value trade-offs would have to be made with MIRVed missiles. In addition, the CMC—on which Secretary Brown's projections regarding throw-weight and counterforce potential were heavily dependent—received a decidedly negative reaction in Congress and elsewhere, primarily because it put too many eggs in one basket. Brown had acknowledged that the CMC would be far more attractive if it were deployed along with a large number of smaller aircraft carrying cruise missiles.[58] Moreover, for some time it appeared that these options would be precluded under SALT II: the deployment of ALCM was to be restricted to "heavy bombers," and the Soviets were insisting that such aircraft be designed and built for that purpose, not modified civil designs. In short, many observers felt the United States was accepting limits on an important new offensive weapon while the Soviet Union remained completely free to deploy new defenses against this weapon.

A second major objection to placing too much faith in the cruise missile was that this weapon, even if proliferated in huge numbers, would not directly address the problem of a major imbalance in first-strike counterforce capability.[59] To a close approximation, in the past and for the foreseeable future, first-strike capabilities are seen to reside in the ICBM forces and here the Soviet lead in throw-weight and number of warheads was expected to increase. Combined with inevitable improvements in accuracy, it was feared the Soviet Union would have an exercisable option to initiate a counterforce first strike, an option that would not be available to the United States. The Carter administration took the position that a large asymmetry in ICBM vulnerability in favor of the Soviet Union was a major concern—though more from the standpoint of perceptions than of a usable military superiority—but not an imminent threat to strategic stability. After all, each of the other two legs of the strategic forces was scheduled to begin significant modernization and expansion before ICBM silo vulnerability reached high levels, even on pessimistic assumptions regarding the effectiveness of future Soviet ICBMs. By the time, or soon after, Minuteman vulnerability reached high levels, the United States would have the option of promptly deploying a new ICBM (MX) in a relatively invulnerable mobile basing mode. As a backup, this direct response to ICBM vulnerability could be brought forward in time by shifting a portion of the Minuteman III force to mobile basing. Further, apart from the daunting technical uncertainties in executing a first strike, two other important considerations allowed Carter administration officials to play down the practical significance of a likely Soviet lead in first-strike counterforce capability. First, Soviet defense planners could never be totally sure the United States would stick to its declared policy of riding out a nuclear attack. The United States might launch its ICBMs once it was convinced a full-scale attack was under way. Second, a full-scale Soviet counterforce attack would almost certainly result in U.S. civilian deaths running into the tens of millions.[60] This being the case, the Soviet Union could never entirely discount the possibility that the immediate U.S. response would be a counterpopulation strike. The alleged irrationality of such a response, to paraphrase Defense Secretary Brown, would be no consolation in retrospect.[61]

As shall be seen in the following chapter, the Carter administration subsequently conceded that the cruise missile initiative—although it could be portrayed as an indirect means of offsetting the imminent Soviet lead in prompt counterforce capability—was not sufficient. In conjunction with the presentation of the FY1980 defense budget, the first steps were taken to ensure that the United States would not appear to be grossly deficient in prompt counterforce capabilities in the 1980s. But as far as the strategic cruise missile was concerned, the die had long been cast.

CONCLUSION

In sum, the long-range cruise missile was assigned a decisive role in the U.S. strategic nuclear posture for the 1980s and, in all probability, for long thereafter. Midway during the development of the various cruise missiles—and after a gap of nearly a year in which, in principle, the United States had no strategic long-range cruise missile[62]—the platform for the strategic weapon was switched from the submarine to the bomber because only the latter could launch the weapon in the requisite numbers.

The evidence that was reviewed earlier suggests strongly that the long-range, submarine-launched cruise missile was never seriously regarded as a major response to the Soviet strategic buildup. During its brief life as a strategic weapons system, the most important role assigned to it was as a reserve weapon. If it is not effectively limited in range after the expiration of the protocol to the SALT II treaty, it will obviously continue to fulfill this role even if, for declaratory purposes, it is assigned a theater mission. That the SLCM was never considered as a major strategic initiative is reflected in the fact that the numbers considered for procurement never seem to have exceeded about 400. It is true that rather much was made of the probability that, if a few SSNs were armed with a few SLCMs each, a prudent opponent, being unable to verify this fact, would assume the worst and regard every SSN as a launch vehicle for 20-plus SLCMs. On the other hand, it was made abundantly clear that the United States had no intention and indeed no real possibility of deploying this many SLCMs on SSNs. It was stressed again and again that the SSN's primary role was and would remain ASW, and that four to six SLCMs per boat was the upper limit if ASW capabilities were not to be compromised.

It was considered that the ALCM would play a major role in keeping the United States ahead in selected indexes of strategic power, and thus satisfy the requirement that U.S. strategic forces be "essentially equivalent" to those of the Soviet Union. Essential equivalence is defined as a condition in which any advantages in force characteristics enjoyed by the Soviet Union are offset by U.S. advantages in other characteristics. As indicated earlier, the ALCM, together with the Trident SLBM program, was expected to preserve a significant U.S. lead in number of warheads through 1986. The ALCM could also preserve the prevailing U.S. lead in hard-target kill potential, but only if deployed on CMCs as well as B-52s.

Beyond this, a force of strategic cruise missiles was less costly, at least over the medium-term future, than other proposed ways of preventing the Soviet Union from assuming the lead on all indexes of strategic power, for example, a large force of new penetrating bombers or a move to mobile ICBMs. Secretary Brown cogently summed up the Carter administration's assessment of the efficacy of the strategic cruise missile as follows:

> I am certain that the cruise missile will improve the world's perception of the potency of our forces, not only by maintaining strategic force parity with the Soviet Union, but also by retaining a clear technological superiority. And . . . we are doing all this with a weapon that because of its long flight time, does not threaten a first-strike capability.[63]

MILITARY REQUIREMENT / 129

NOTES

1. The magnitude of the asymmetry was in fact significantly overstated for a time. From 1968 through 1975 the missile cannisters observed on the Kresta II and Kara-class cruiser and Krivak-class destroyers were believed to contain a 30 n.m. antiship weapon designated SSN-10. In 1975 these three classes numbered 18 vessels carrying a total of 108 launchers for the SSN-10. Early in 1976 it was discreetly acknowledged that the SSN-10 was in fact an antisubmarine weapon with the new designation SSN-14.

2. This was made clear in 1972 by the then CNO, Admiral Zumwalt. See Fiscal Year 1973 Authorization for Military Procurement, hearings, p. 994.

3. This point was made on several occasions. For example, see Fiscal Year 1976 and July-September Transitional Period Authorization for Military Procurement, hearings, p. 5,154.

4. For example, Jane's All the World's Fighting Ships 1972/73.

5. A range of about 470 n.m. can be deduced from a map of the United States showing the SSN-3's target coverage (see Fiscal Year 1974 Authorization for Military Procurement, hearings, p. 2,633). In 1976 Aviation Week and Space Technology (2 February 1976, p. 12) credited the latest surface-ship version of the SSN-3 with a range of 550 n.m. In a much-deleted exchange on the range of Soviet cruise missiles in congressional hearings in February 1977, a figure of 700 miles was given (Hearings on Military Posture and HR5068 [HR 5970], House Armed Services Committee, February-March 1977, p. 1,110).

6. Aviation Week and Space Technology, 2 February 1976, p. 12. Interestingly enough, rumors of a new Soviet submarine-launched cruise missile with a range comparable to that proposed for the SLCM emerged as early as July 1972 (Space Business Daily, 31 July 1972, p. 149).

7. These range figures are derived from a graph inserted in Hearings on Military Posture and HR 1150, House Armed Services Committee, Part 5, February 1976, p. 193. The 2,000 n.m. range attributed to the SLCM would put over 90 percent of the Soviet population at risk on the same assumptions regarding standoff.

8. Aviation Week and Space Technology (31 March 1975, p. 13) claims that the first version of the SSN-3 tested (which would be back in the latter half of the 1950s) was a strategic configuration.

9. Recently the chairman of the Joint Chiefs of Staff stated that "it appears clear that a significant portion of the Soviet naval cruise missile inventory is nuclear armed" (United States Military Posture for FY1979, p. 9).

10. Fiscal Year 1976 and July-September Transitional Period Authorization for Military Procurement, hearings, p. 5,208. It was

indicated elsewhere that the nuclear warhead on an unspecified Soviet cruise missile (but probably the SSN-3 or SSN-12) was in the kiloton range. Since the large majority of U.S. strategic nuclear warheads have a yield of 200 KT or less, there is probably an inclination to regard any long-range weapon with a warhead of this yield or greater as strategically rather than tactically oriented. On the other hand, U.S. Navy data indicate that a ship will sink if an attacking nuclear warhead generates an overpressure of 30 to 50 psi. A one-MT warhead could sink a ship if it landed one mile away. A 130-KT warhead would suffice if accuracy is doubled to half a mile. In other words, a presumption that Soviet cruise missiles have a strategic rather than a tactical role that derives from the size of the nuclear warhead depends critically on the accuracy of the weapon in the tactical role. If it is relatively inaccurate a very large warhead would not be inconsistent with a tactical role.

11. Fiscal Year 1974 Authorization for Military Procurement, hearings, p. 2,631.

12. Fiscal Year 1975 Authorization for Military Procurement, hearings, p. 3,621 (emphasis added).

13. Fiscal Year 1976 and July-September Transitional Period Authorization for Military Procurement, hearings, p. 5,129.

14. Statement by Navy Secretary W. Graham Claytor reported in International Herald Tribune, 17-18 June 1978, p. 3.

15. Department of Defense Appropriations for 1977, hearings, House Committee on Appropriations, Part 2, February 1976, p. 172.

16. D. J. Ball, "The Strategic Missile Programme of the Kennedy Administration," Ph.D. diss., ANU, June 1972, p. 2.

17. Lt. Gen. Daniel O. Graham, "The Intelligence Mythology of Washington," Strategic Review (Summer 1976): 62.

18. SIPRI Yearbook of World Armaments and Disarmament, 1968/69 (Stockholm: Almqvist and Wiksell, 1969), p. 33, and Melvin R. Laird, secretary of defense, Annual Defense Department Report, FY1973, 15 February 1972, p. 36.

19. James R. Schlesinger, secretary of defense, Annual Defense Department Report, FY1975, 4 March 1974, p. 45.

20. Ibid., p. 46.

21. Ibid., pp. 46-47.

22. SIPRI has cited "senior U.S. defense officials" as stating that a "production run" of SS-16 missiles had been stockpiled but not deployed (World Armaments and Disarmament, SIPRI Yearbook 1976 [Stockholm: Almqvist and Wiksell, 1976], p. 188).

23. Defense Monitor, Washington, D.C.: Center for Defense Information, September-October 1978.

24. Harold Brown, secretary of defense, Department of Defense Annual Report, Fiscal Year 1980, 25 January 1979, p. 72. Through

January 1979, the SSN-X-17 had been backfitted into just one Yankee-class submarine and, as indicated by the "X," was still considered to be under development.

25. Defense Monitor, September-October 1978.

26. General George S. Brown, chairman of the Joint Chiefs of Staff, United States Military Posture for FY1979, 20 January 1978, p. 31.

27. The first is Colin S. Gray, "Soviet Rocket Forces: Military Capacity, Political Utility," Air Force Magazine (March 1978): 52, and the second is from a table prepared by the Library of Congress and reprinted in Aviation Week and Space Technology, 18 April 1977, p. 18.

28. Aviation Week and Space Technology, 5 December 1977, p. 12.

29. Gray, "Soviet Rocket Forces," p. 52.

30. Donald H. Rumsfeld, secretary of defense, Annual Defense Department Report FY1978, 17 January 1977, p. 61.

31. Ibid.

32. Congressional testimony by Defense Secretary Harold Brown reported in Aviation Week and Space Technology 8 August 1977, p. 15.

33. It must be pointed out, however, that one official U.S. publication claims that the United States led the Soviet Union in SLBM throw-weight in 1978. This contradicts the data here. See The Strategic Arms Limitations Talks, Department of State, Special Report No. 46, July 1978, p. 12. The probable explanation is that the throw-weight of the Poseidon missile is considerably larger than the 2,200 to 2,500 lbs. commonly assumed. It is considered technically infeasible to reduce the weight of a ballistic reentry vehicle below about 300 lbs., and current models will be somewhat heavier. Thus even if the Poseidon carries only eight MIRVs, its throw-weight is probably well over 3,000 lbs. since allowance must also be made for the weight of the guidance system and the MIRV bus.

34. Three of these are: Kosta Tsipis, Offensive Missiles, SIPRI: Stockholm Paper 5, 1974; Edward Luttwak, The US-USSR Nuclear Weapons Balance, Washington Papers, No. 13, Washington, D.C.: Center for Strategic and International Studies, 1974; and Congressman Robert L. Leggett, "Two Legs Do Not a Centipede Make," Armed Forces Journal International (February 1975): 20-32. All three have been succinctly summarized and evaluated by Thomas A. Brown, "Missile Accuracy and Strategic Lethality," Survival (March-April 1976): 52-59.

35. An excellent discussion of the fratricide problem is Joseph J. McGlinchey and Jacob W. Seelig's, "Why ICBMs Can Survive a Nuclear Attack," Air Force Magazine (September 1974): 82-85.

36. *Aviation Week and Space Technology*, 8 August 1977, p. 15.
37. Fred A. Payne, "The Strategic Nuclear Balance: A New Measure," *Survival* (May–June 1977): 103-10.
38. United States acceptance of the SS-19 was a notable (but legally necessary) concession. The clause in the SALT I agreement that limits the permissible increase in volume of a replacement missile was, after much confusion, unilaterally interpreted by the United States to mean that the "maximum volumetric increase in missile size permitted cannot exceed approximately 32 per cent" (*Military Implications of the Strategic Arms Limitation Talks Agreements*, hearings, House Armed Services Committee, June–July 1972, pp. 15, 144–45).
39. *Department of Defense Annual Report, Fiscal Year 1979*, 2 February 1978, p. 5.
40. Ibid., p. 63. A similar warning had been given a year earlier by Brown's predecessor, Donald Rumsfeld.
41. Ibid., p. 64.
42. Colin S. Gray, "SALT II and the Strategic Balance," *British Journal of International Studies* (1975): 201.
43. John D. Steinbruner and Thomas M. Garwin, "Strategic Vulnerability: The Balance between Prudence and Paranoia," *International Security* (Summer 1976): 140. For a broader statement of the thesis that grave concern about a destabilizing first-strike threat is unjustified, at least on the basis of rational military considerations, see Donald R. Westervelt, "The Essence of Armed Futility," *Orbis* (Fall 1974): 689-705.
44. A. A. Tinajero, *Projected Strategic Offensive Inventories of the U.S. and USSR*, Congressional Research Service, Library of Congress, 24 March 1977.
45. Remarks by Paul Nitze cited in *Aviation Week and Space Technology*, 5 December 1977, p. 12.
46. Paul H. Nitze, "Deterring our Deterrent," *Foreign Policy* (Winter 1976-77): 195-210. Earlier, Nitze had presented a lengthier but less quantitative statement of his views. See his "Assuring Strategic Stability in an Era of Detente," *Foreign Affairs* (January 1976): 207-32.
47. Nitze points out that the calculations were done by T. K. Jones, his technical advisor while he was on the U.S. SALT delegation, on the basis of classified data (Nitze, "Assuring Strategic Stability," p. 225).
48. For a detailed rebuttal of Nitze's arguments, see Jan M. Lodal, "Assuring Strategic Stability: An Alternative View," *Foreign Affairs* (April 1976): 462-81.
49. *Department of Defense Annual Report, Fiscal Year 1979*, p. 63.

50. The latter part of this sentence is taken from Colin S. Gray, "The Strategic Forces Triad: End of the Road?" Foreign Affairs (July 1978): 774.

51. Annual Defense Department Report, Fiscal Year 1979, p. 104.

52. See, for example, Des Ball, "The Counterforce Potential of American SLBM Systems," Journal of Peace Research (1977): 23-40, and K. Tsipis, A. H. Cahn, and B. T. Feld, eds., The Future of the Sea-Based Deterrent (Cambridge, Mass.: MIT Press, 1973).

53. The fleet ballistic missile accuracy improvement program has been a standard request in the U.S. Navy budget for many years.

54. Fiscal Year 1977 Authorization for Military Procurement, hearings, Senate Armed Services Committee, February-March 1976, p. 3,368.

55. Summarized in Aviation Week and Space Technology, 8 August 1977, p. 15.

56. Assuming a warhead yield of 170 KT and a CEP of 300 feet, an ALCM would destroy targets hardened to 1,000 psi and 2,000 psi with a probability of 99.5 percent and 97.5 percent, respectively. Thus the arrival of around 40 percent of the projected force of 3,400 ALCMs could essentially destroy all Soviet ICBM silos.

57. This suggests that the analyst, Peter Hughes, was even more optimistic on the survivability of the cruise missile than Secretary Brown. According to Brown, with the B-52/ALCM force alone, the Soviet Union would have a superiority in counterforce potential of 1.5 times that of the United States in 1986 before a Soviet first strike.

58. Aviation Week and Space Technology, 8 August 1977, p. 15.

59. For a forceful statement of this point of view, see Gray, "The Strategic Triad: End of the Road?," pp. 781-84.

60. Early in 1974, the then secretary of defense, James Schlesinger, remarked that a Soviet attack on U.S. ICBMs could, under certain conditions, result in fewer than one million fatalities. The remark was made in the context of Schlesinger's campaign to persuade U.S. decisionmakers to entertain the prospect of and to plan for nuclear war scenarios in which any massive response, whether counterforce or countervalue, might be quite inappropriate. Subsequent studies concluded that a Soviet attempt to destroy the U.S. ICBM force would kill anything up to 22 million Americans. The toll would go even higher if the attack were also directed against bomber bases and SSBN ports (see Effects of Limited Nuclear War, hearings, Subcommittee on Arms Control, International Organizations and Security Agreements of the Committee on Foreign Relations, U.S. Senate, 18 September 1975).

61. Department of Defense Annual Report, Fiscal Year 1979, p. 63.

62. That is between the time in 1976 when the SLCM was reoriented toward a theater role, and January 1977 when the Air Force was instructed to give first priority to a long-range standoff version of the ALCM.

63. <u>Department of Defense Annual Report, Fiscal Year 1979</u>, pp. 115-16.

6

STRATEGIC DOCTRINE: COUNTERFORCE, LIMITED OPTIONS, AND THE CRUISE MISSILE

INTRODUCTION

Since the dawn of the nuclear age, nuclear weapons have presented the most profound dilemma. The destructive capacity of these weapons was so immense, so much in excess of anything previously experienced, that few people could conceive of circumstances in which the use of these weapons would constitute a rational act. If both warring parties possessed large stockpiles of nuclear weapons their employment, to all intents and purposes, was tantamount to national suicide. No national interest or aggregation of interests could be held to be so vital as to be commensurate with this risk.

Nevertheless, from 1949 two mutually antagonistic powers possessed nuclear weapons and considerations had to be given to the principles that should govern the size and structure of the nuclear forces and what one should declare to be the purpose of these forces. Reconciling the certainty of mutual devastation in nuclear war with the facts of great power competition and the role of force in international affairs has been a Gordian knot that has provoked ceaseless debate. In the United States the official position on the manner in which and under what circumstances strategic nuclear weapons would be employed has periodically shifted quite markedly. The latest major shift took place while the long-range cruise missile program was in its infancy. The purpose of this chapter is to determine whether or not there was any significant interrelationship between this development in strategic doctrine and the initiation and subsequent vicissitudes of the cruise missile programs.

The relationship between strategy and weapons technology is a complex one with causality running both ways. In the United States, shifts in declared strategic policy have been driven at least as much by developments in nuclear weapons technology as the other way around. A similar imperative is almost certainly at work in the Soviet

Union, although this country is less forthcoming on how it views the purpose of strategic nuclear weapons.[1] Despite this, it remains the case that changes in the official view on how the size and structure of the nuclear forces should be determined, and related changes in declared policy regarding the use of these forces, do accelerate or retard the development of particular new weapons or the development of particular capabilities for existing weapons.

Although only the most recent restatement of strategic nuclear policy in the United States is of immediate interest to this study, it is important to stress the evolutionary nature of revisions in this policy. United States strategic policy has never moved abruptly nor completely from one basis to another; all the changes have been changes in degree. Significantly, however, the changes have predominantly been in one direction: to place greater emphasis on the ability to wage war with strategic nuclear weapons.

President Kennedy and Defense Secretary Robert McNamara sought to disaggregate strategic policy from the single massive retaliatory attack the Eisenhower administration had maintained would be the U.S. response to a threat to its vital interests. In a radio and television address in July 1961, Kennedy stressed that the United States intended to "have a wider choice than humiliation or all-out nuclear action."[2] This statement had implications for both conventional and nuclear forces. Kennedy administration officials took exception to the implication of the doctrine of massive retaliation that nuclear weapons could substitute for conventional forces. They were more concerned to stress the "firebreak" between conventional and nuclear weapons and therefore sought, both for the United States and its NATO allies, greater conventional capabilities in order to push the nuclear option higher up the scale of potential conflict.

In the nuclear arena, the quest for a wider choice emerged as the doctrine of controlled response. In public elaborations of this doctrine—notably McNamara's speech in Ann Arbor, Michigan, in June 1962—it was made clear that controlled response could be equated with damage limitation. The United States would seek the ability, in response to a Soviet nuclear attack, to destroy residual Soviet nuclear forces in order to limit the scale of a second Soviet attack and, perhaps, to terminate the fighting before it escalated to direct attacks on population centers. Because of this latter element, the policy of controlled response also became known as the "no-cities" doctrine. This damage limitation capability—calling for accurate and quickly retargetable weapons under high quality command and control—would be additional to the forces required to inflict massive damage on Soviet population and industry or, in McNamara's terminology, the forces for assured destruction.

As far as declaratory policy was concerned, the prominence

assigned to damage limitation was to be short-lived. By 1963 McNamara and other officials began to stress the unpredictability of the course a nuclear exchange would take and to point out that, while damage limitation was feasible at the time, it would become infeasible as the Soviet Union acquired more numerous, dispersed, and hardened strategic forces. Over the remaining five years of his term as secretary of defense, McNamara placed increasingly exclusive emphasis on assured destruction as the yardstick for measuring the adequacy of U.S. strategic forces. Assured destruction was particularly amenable to the analytic techniques favored by McNamara for determining force posture. The only unscientific step required was to judge the degree of destruction necessary to deter the Soviet Union from initiating a nuclear war. As is well known, McNamara judged that the potential to kill 25 to 30 percent of the Soviet population and destroy one half to two thirds of its industrial capacity would suffice as a deterrent. Even here technical factors appear to have intruded: it happens that beyond these levels of destruction, the additional damage per additional warhead is subject to sharply diminishing returns.

One could hardly argue that the emphasis on assured destruction stunted the development of U.S. strategic capabilities; the employment of greater-than-expected threat and worst-case scenarios rationalized a very vigorous program indeed. Nevertheless, it is clear that McNamara deliberately sought to play down, and eventually to exclude, the need for counterforce or damage-limiting capabilities because of his mounting concern over the internal dynamics of the strategic arms race. Technical developments would lead to enhanced counterforce capabilities irrespective of whether U.S. strategic policy required this. To make damage limitation a formal component of the computations used to determine force posture would make the bureaucratic pressures to accelerate the acquisition of counterforce capacity that much more difficult to resist.

McNamara's increasingly intensive efforts to dampen the growth of U.S. strategic offensive capabilities met with very limited success, a tribute to the strength of the technological and bureaucratic pressures resisting restraint. It is quite clear that the targeting of nuclear weapons for purposes of damage limitation remained in the SIOP since 1962 despite the rise of the assured destruction policy. When a senator in early 1968 asked the chairman of the Joint Chiefs of Staff, General Wheeler, whether U.S. "war plans allocate weapons for damage limitation or counterforce," Wheeler replied, "They certainly do."[3] At these same hearings, which took place shortly after McNamara's departure, Wheeler could complain only that emphasis on assured destruction had inhibited the accumulation of damage-limiting capabilities, not that it had prevented it altogether.[4] Finally, perhaps the clearest indication that damage limitation lingered on as a factor shaping the size and targeting plans of U.S. strategic forces was provided by Harold Brown when he was secretary of the Air Force. Early in 1968 he had this to say:

> we strive to maintain a capability that will appear visibly unassailable to a rational man, but that will also permit us, if reason and deterrence should fail, to use these forces, to the extent feasible, to neutralize opposing weapons and limit damage to ourselves and allies.[5]

After McNamara's departure, a widely supported effort was mounted to diversify the criteria used to gauge the adequacy of U.S. strategic forces. The fact that the Soviet Union was rapidly approaching parity in numbers of ICBMs and, one suspects, the feeling that the United States should not enter the SALT negotiations without a thorough review of what was expected of the strategic forces, probably lent an element of urgency to this process. In particular, this review had to contend with the widely held view that the United States could and should retain clear strategic superiority, a notion McNamara had increasingly viewed as empty of meaning.[6]

During 1969 an interagency review of military policy, including strategic policy, was conducted.[7] In 1970 President Nixon both encouraged and foreshadowed a change in policy when he essentially repeated the concern Kennedy had expressed ten years earlier:

> Should a President, in the event of an attack, be left with the single option of ordering the mass destruction of enemy civilians in the face of the certainty that it would be followed by the mass slaughter of Americans? Should the concept of assured destruction be narrowly defined and should it be the only measure of our ability to deter the variety of threats we may face?[8]

The results of this reexamination emerged in 1971 under the label of "sufficiency."[9] Strategic sufficiency involved meeting the following four criteria:

(a) maintaining an adequate second-strike capability to deter an all-out surprise attack on U.S. strategic forces;
(b) providing no incentive for the Soviet Union to strike the United States first in a crisis;
(c) preventing the Soviet Union from gaining the ability to cause considerably greater urban/industrial

destruction than the United States could inflict on the Soviets in nuclear war; and

(d) defending against damage from small attacks or accidental launches.[10]

This list did not represent a major departure from the objectives established in the past, but it did contain the seeds of change. Assured destruction remained the cornerstone of strategic doctrine, but it was not held to be equivalent to sufficiency. In particular, criteria (b) and (c) allowed ample scope for the acquisition of counterforce capabilities; (b) required that the United States, in a retaliatory counterforce strike, be able to eliminate any imbalance created by a Soviet first strike, and (c) required that the countervalue potential of U.S. forces remaining be at least as great as that of the Soviet forces remaining. In other words, the amount of assured destruction required would be measured on a relative scale rather than on the basis of some arbitrary absolute requirement. Also, the desire to reduce Soviet incentives for a first strike had not previously been an explicit element of planning, although this is not very different from the long-established requirement that strategic forces be invulnerable.

In addition to these itemized components of the doctrine of sufficiency, Nixon administration officials frequently referred to the notion of flexibility as another criterion for gauging the sufficiency of strategic forces. This criterion was not discussed in any detail and no major campaign was mounted to gather support for a decisive and open move toward flexible options for the strategic forces. Nevertheless, it was evident that the administration regarded the strategic war plans inherited from the Johnson era as too rigid and providing too few options for the employment of strategic weapons.

Even the administration's expressions of interest in flexible options proved controversial. In particular, Congress demonstrated it was not prepared to endorse programs that seemed wholly or primarily aimed at the acquisition of greater hard-target kill capabilities. Late in 1969 congressional pressure forced the administration to cancel a development program to improve the accuracy of Poseidon warheads. Similarly, a hard-target warhead project proposed in the SALT-I–related supplementary budget was dismissed by Congress primarily on the ground that no technical distinction existed between a first-strike and a second-strike hard-target weapon, and that any movement toward the former would be destabilizing. These experiences presumably convinced the administration that acceptance of the principle of flexible options (including some counterforce options) would require: a deliberate effort to discredit the assured destruction strategy; and the presentation as an alternative of a well-prepared, theoretically sound package that gave high priority to flexibility.[11]

These official efforts to move beyond assured destruction—or minimal deterrence as it is also known—were complemented by mounting criticism of the concept in the strategic community at large. Many observers questioned the psychological validity of the theory, particularly the requirement that leaders of nuclear weapon states be rational. Apart from the prospect of a maniac at the head of a nuclear power, the behavior of a normally rational leader during a crisis could not be assumed always to result in actions designed to avoid war. Furthermore, communications between leaders during a crisis may not convey the intended message, and signals in the form of alerts or force deployments may be misinterpreted.

A second line of criticism was directed at the international political effects of an assured destruction deterrent. The notion of an assured destruction deterrent had the effect of isolating the strategic forces of the United States and the Soviet Union. For the United States, the perception during the 1960s that this was happening generated a great deal of apprehension in its NATO allies. Ostensibly, U.S. strategic forces are linked, through tactical and theater nuclear weapons, to the defense of Western Europe. The emphasis on assured destruction—and increasingly mutual assured destruction—for the deterrence of nuclear war made many U.S. allies doubt the credibility of any declarations that provided for the use or threat of use of U.S. strategic forces in any situation other than a direct Soviet attack on the United States. In short, if deterrence based on assured destruction failed, it left no roads open other than total holocaust. If it worked it left allies feeling vulnerable.

A third criticism, and perhaps the most influential one, was based on ethics. One of the strongest ethical principles of war is that noncombatants should be spared. However, the doctrine of assured destruction is a promise to attack civilians directly with the largest and most destructive weapons ever conceived. In terms of traditional ethics, a counterpopulation deterrent is patently immoral. "To put the point bluntly, if counter-population warfare is murder, then counter-population deterrent threats are murderous."[12] An unethical deterrent strategy is bad per se; but the very fact that it is unethical may cause the deterrent to fail because an opponent may judge that, in the final analysis, its enemy will not carry out the promised retaliation.

For these various reasons many people argued that while assured destruction may have been a necessary doctrine in the 1950s and early 1960s, when nuclear weapons were generally large, dirty, and relatively inaccurate, the technical capability now existed to move toward a less inhuman strategy for the deterrence of nuclear war.[13]

On the other hand there are those who maintain that nuclear weapons are unique. Any notions that one can use them in a manner analogous to conventional weapons are naive and, insofar as these notions might make the decision to use nuclear weapons easier to take, exceedingly dangerous. To paraphrase one writer, so long as the nuclear arsenals remain at anything like their present levels, the mutual hostage relationship between the citizens of the United States and the Soviet Union, though far from ideal, will remain a matter of physical fact and quite insensitive to changes in strategic policy.[14]

In this climate of general dissatisfaction with a deterrent based on the threat of mass annihilation, but with very different opinions on whether it was possible or desirable to do anything about it, Defense Secretary James Schlesinger undertook to publicly commit the United States to a revised nuclear strategy. This strategy provided for a wide range of options for the employment of strategic nuclear weapons, and left assured destruction as a reserve option to be used only as a last resort.

It is possible to identify a number of factors that had some direct bearing on the timing of the move to revise U.S. declaratory policy on the employment of nuclear weapons. Perhaps the most important of these was the fact that the president, his national security advisor, and his secretary of defense were, on this subject at least, of the same opinion. President Nixon, while no strategist, had repeatedly indicated his desire for a greater diversity of options for the use or threat of use of strategic nuclear weapons. Similarly, Henry Kissinger, in his two major academic deliberations on the subject, stressed the desirability of "alternatives less cataclysmic than thermonuclear holocaust," or a "spectrum of capabilities," and for nuclear weapons that are "flexible and discriminating."[15] Finally, Schlesinger had summarized his views in a short paper published in 1968 while he was an analyst at the Rand Corporation:[16]

> with the buildup of major devastation potential on both sides, it becomes increasingly doubtful whether such capabilities are the most reassuring or desirable to possess or whether they could serve any useful purpose in that range of contingencies which are becoming increasingly probable (p. 9).
>
> In the event of war the most desirable thing that can happen to weapons of mass destruction is that they be destroyed before they inflict damage (p. 17).
>
> If war can be kept at a low level and directed toward military rather than urban targets, it would seem to me to be most consistent with the objectives of arms control (p. 18).

A second instigating factor was the Soviet testing of MIRVed ICBMs, begun in August 1973. Though hardly unexpected, this event fueled predictions that the Soviet Union fully intended to exploit its advantage in numbers and throw-weight of missiles to secure a major lead in counterforce potential. Others saw it as a major step toward providing the capabilities necessary to support Soviet nuclear strategy, which is a war-fighting strategy.

One of the by-products of the negotiations on the limitation of strategic arms was an intensified effort to discover how the Soviet Union viewed the purpose and utility of strategic nuclear weapons. Many Americans felt that a prerequisite for the success of SALT would be a meeting of the minds on this issue. Accordingly, it was openly acknowledged that, if it proved necessary, the first order of business at SALT would be to educate the Soviets in the logic of deterrence—specifically, that neither side should do anything to diminish its vulnerability to a nuclear strike by the other. The 1972 treaty limiting ballistic missile defense complexes to two sites (subsequently reduced to one) was regarded as clear evidence that the two sides had come to a common understanding on the role of strategic weapons.

Since then, however, analyses of Soviet writings on military strategy, the direction and scope of their strategic weapon development and deployment program, and the magnitude of their efforts in the area of civil defense have raised serious doubts about the commitment of the Soviets to mutual vulnerability as the only sensible basis for the strategic relationship. The fact of the ABM treaty can be explained by arguing that the Soviet Union feared a U.S. system would be far more effective than its own and sought time to overcome the technological lag. The fact that the Soviet Union has maintained an intensive R&D effort on ABMs can be cited to support this theme, although it must be borne in mind that the same is true of the United States. A variation on this theme is that if the United States deployed effective defenses against ballistic missiles, and if it dedicated these systems to the defense of offensive weapons rather than population centers, then this would directly threaten the Soviet deterrent.

SOVIET STRATEGIC DOCTRINE

Piecing together a coherent and credible picture of Soviet nuclear strategy is no easy task for Western analysts. The Soviet style of writing, the manner of presentation, the heavy ideological undertones, and the virtually total absence of hard data (except when reference is made to Western military forces) make the Soviet source material quite alien to the Westerner accustomed to the wealth of information contained in posture statements, congressional hearings, trade journals, and the like. Nevertheless, a substantial body of literature on the subject exists. On the whole, this literature leaves no doubt that Soviet declaratory nuclear strategy is markedly different from that of the United States, particularly before the appointment of Schlesinger as secretary of defense.

A useful point of departure is that, in the Soviet Union, strategy and tactics are determined far more exclusively by the professional military than is the case in the United States. Few dispute that the military is under effective civilian control, but the U.S. practice of filling top military posts with civilians is not followed in the Soviet Union. In addition, while the United States has a large and influential community of civilian specialists in military affairs, its counterpart in the Soviet Union is small, more recent, and much less influential. As a result, Soviet nuclear strategy is shaped predominantly by traditional military concerns and considerations. Given this, it is scarcely surprising that the notion that the deterrence of nuclear war should rest on maximum mutual vulnerability to nuclear attack is simply not entertained in Soviet discourses on the subject.

To the contrary, Soviet commentary stresses the fact that the possibility of nuclear war must be squarely faced: "The monstrous growth of the means of destruction does not make a world nuclear war automatically impossible."[17] Once this step is taken it follows inevitably that one should endeavor, first, to survive and, second, to prevail in such a war. At the most fundamental level, strategic nuclear war is held to be no different from any other kind of war:

> The premise of Marxism-Leninism on war as a continuation of policy by military means remains true in an atmosphere of fundamental changes in military matters. The attempt of certain bourgeois ideologists to prove that nuclear missile weapons leave war outside the framework of policy and that nuclear war moves beyond the control of policy, ceases to be an instrument of policy and does not constitute its continuation is theoretically incorrect and politically reactionary.[18]

The predictable military response to this kind of directive is to prepare to defend the country in the event of nuclear war. Soviet military strategists, like their U.S. counterparts, have concluded that the "rivalry between the means of attack and the means of defense at present is characterized by the superiority of the former, due to the present development level of nuclear missile weapons."[19] Similarly, the offense being dominant, the importance of surprise is greatly enhanced.[20] The predominant U.S. response to this state of affairs has been to distinguish between deterrence of and defense against nuclear attack, and to concentrate heavily on the former; that is, to insure that even in the face of the most massive and brilliantly executed Soviet surprise attack, adequate nuclear forces would survive to inflict an unacceptable degree of damage on Soviet population and industry.

The Soviet response, in contrast, does not reflect any distinction between deterrence and defense. The objective is the same: to convince the opponent that "the price of aggression comes too great and does not justify those goals for which it is undertaken."[21] But the means are more straightforwardly military-technical: to show a willingness to wage nuclear war and to aspire to the various capabilities required. The following quotation elaborates on the scenario that appears to dominate Soviet strategic planning:

> The implicit scenario for the Soviets requires successful anticipation of an imminent U.S. "surprise attack." Thus, the strategic forces of the United States, assuming sufficient warning of the impending American attack would be largely destroyed by a preemptive strike. Those U.S. forces which survived Soviet preemption and were actually launched would be met by the massive Soviet air defenses, and greatly degraded. Those, finally, which succeeded in delivering their weapons to their targets would have attacked a population effectively organized and, to the degree feasible, protected by a vigorous civil defense program and an economy and political control structure also organized to cope with such an attack.[22]

This, of course, is a highly condensed formulation of the strategic doctrine one can distill from Soviet literature on the subject, but in general terms there is little dissent from it.

The question that deeply divides Western analysts is whether or not it follows that the Soviet Union believes strategic superiority to be an attainable objective and, once attained, exploitable for political advantage. Put another way, is the Soviet Union less terrified of the prospect of strategic nuclear war than is the United States?

One side of the debate argues that Soviet authorities, including military authorities, are clearly just as aware as their U.S. counterparts that nuclear war would be catastrophic. For example:

> a nuclear war, if the imperialists are able to start it, will lead to a gigantic waste and to the direct destruction of a significant portion of the productive forces, including the main productive force, the working man. A war can cause

enormous damage to the very fundamentals of society's existence, to social progress and all world civilization.[23]

The argument proceeds that, as in the United States, deterrence is the primary mission of the Soviet strategic forces, but that for military doctrinal and related historical and ideological reasons the operating definition of strategic deterrence is quite distinct from the U.S. one. The military doctrinal factor and the resulting emphasis on the war-fighting capacity of weaponry have already been encountered. The intense preoccupation with defense was created by the Russian-Soviet experience of devastating and just barely repulsed invasions and the long-standing perception of being surrounded by enemies. Finally, given the fundamental tenet of Marxist-Leninist ideology that the victory of socialism is inevitable, Soviet authorities simply cannot openly acknowledge the very real possibility that a U.S. nuclear attack would stop Soviet socialism dead in its tracks. In sum, "the Soviet view appears to be that the better their armed forces are prepared to fight and win a nuclear war, and the more any adversary knows this to be the case, the more successful is Soviet deterrence."[24]

The other side of the debate hinges its argument to a large extent on the Soviet civil defense program. United States interest in the implications of the Soviet civil defense effort was revived in the early 1970s, probably by the publication of a book by Leon Gouré, a U.S. specialist in this field.[25] There is little doubt that the Soviet Union devotes more resources to civil defense than does the United States and has done so for many years. As is the case for other dimensions of the Soviet military program, the expenditure estimates circulating in the U.S. literature are subject to a wide margin of error.

It is also the case that Soviet authorities publicly assign a far more important role to civil defense than do their U.S. counterparts. The following quotation is fairly typical:

> the role of civil defence has grown immeasurably, and its functions are organically intertwined in the process of military operations which can cover the entire territory of the nation.[26]

The civil defense issue came to a head in 1976-77 with a group of influential analysts mounting a determined campaign to warn U.S. officials that the Soviet Union did not subscribe very strongly to deterrence. On the contrary, it was determined to acquire the capabilities necessary to fight and win a strategic nuclear war and, indeed, was well down the road toward this objective. In 1976 a team of analysts under the direction of Thomas K. Jones, working under the auspices of the Boeing Aerospace Company, evaluated the effectiveness of

Soviet civil defense measures in terms of protecting both population and industry.[27] Among the conclusions of this study were: a U.S. retaliatory second strike directed at the Soviet population (assuming the Soviets had three days to implement their civil defense measures) would result in 10 to 11 million casualties; and "if the observed examples of industrial facility dispersal and separation become the pattern for a significant portion of the Soviet Union's future capital expansion, their industry would require little or no preattack hardening to survive and recover rapidly from a nuclear war."[28]

Also in 1976, Gouré published a more extensive volume on the Soviet civil defense program and what it implied for the United States.[29] In response to a critique of this volume, Gouré reiterated his belief "that the Soviet civil defence program poses a serious challenge to the U.S. deterrence posture and retaliatory strategy."[30] Finally, perhaps the strongest statement to the effect that the Soviet Union had a war-fighting and war-winning strategic doctrine came from Richard Pipes in an article in <u>Commentary</u> in July 1977.[31] Pipes argued, citing the Boeing report referred to above, that "the notion underpinning our mutual assured destruction doctrine since the days of McNamara, namely that our second-strike deterrent would destroy one-quarter of the Soviet population and two-thirds of Soviet industry, seems to be quite irrelevant in the light of the facts as known today."[32] He further argued: "we can say with a high degree of confidence that [the Politburo] believes that should nuclear war break out, for whatever reasons, the Soviet Union would be able to fight, survive and emerge meaningfully victorious from it."[33]

After this the controversy was quite effectively defused, at least for a time. The Carter administration undertook a wide-ranging review of global power relationships known as Presidential Review Memorandum 10, completed in June 1977. One input to this review, a document entitled "Military Strategy and Force Posture Review," concluded that, at a minimum, the United States and the Soviet Union would lose 140 and 113 million people respectively in a major nuclear war; with losses on this scale, "neither side could conceivably be described as a winner."[34] A CIA study, released in July 1978, reaffirmed this assessment.[35]

Nevertheless, the civil defense controversy was not without impact. Secretary of Defense Brown has expressed concern over the psychological impact a vigorous and comprehensive civil defense effort may have on Soviet political leaders. In his words, "they may—mistakenly, in my belief—arrive at the conclusion that they could survive as a functioning and powerful country after an all-out thermonuclear exchange."[36] In his first annual report, Brown essentially reiterated this concern:

Neither Minuteman vulnerability nor Soviet civil defense on the scale we now see can seriously degrade our basic retaliatory response. But we must be concerned about perceptions of Soviet superiority based on these two factors.[37]

Finally, President Carter's proposed reorganization of the U.S. civil defense program, a plan submitted to Congress in June 1978, is considered by some to be a first step toward matching the Soviet effort.[38]

This brief review of the Soviet approach to security in the nuclear era and the controversy over what this approach implies for Soviet intentions has, in chronological terms, gone well beyond the terms of James Schlesinger as secretary of defense. It does, however, substantiate the earlier point that the growing realization that the Soviet Union was committed to its own, unique, strategic doctrine, and that the ambitious ICBM and MIRV development programs revealed in 1973 potentially represented a major step forward in implementing that strategy, was an important factor in motivating Schlesinger to revise U.S. strategic policy.

Third, this review, together with the material surveyed in the previous chapter, suggests the following generalization: since the mid-1960s the United States possessed a strategic arsenal relatively suited to war fighting and, until 1974, declared its policy to be one of pure deterrence, while the Soviet Union possessed a deterrent arsenal and subscribed to a war-fighting strategy.

THE SCHLESINGER DOCTRINE

As indicated earlier, the centerpiece of the Schlesinger doctrine was that the United States should undertake the necessary planning to secure the availability of a range of limited strategic options well below the scale of an assured destruction attack.[39] Schlesinger opened his case with the argument:

> that it is only in the process of examining why and how deterrence might fail that we can judge the adequacy of plans and programs for deterrence. And once that analysis begins, it quickly becomes evident that there are many ways, other than a massive surprise attack, in which an enemy might be tempted to use, or threaten to use, his strategic forces to gain a major advantage or concession. It follows that our own strategic forces and doctrine must

take a wide range of possibilities into account if they are successfully to perform their deterrent functions.[40]

He went on to identify four situations in which U.S. strategic forces were expected to have a deterrent effect, either on their own or in conjunction with conventional forces.[41] Specifically, the strategic forces were expected:

> To forestall direct attacks on the United States;
> To deter nuclear attacks on U.S. allies;
> To have a deterrent effect against massive nonnuclear assaults, although the primary responsibility here rested with U.S. and allied theater forces; and
> To inhibit coercion of the United States and its allies by nuclear powers.

In order to perform these deterrent functions, Schlesinger outlined four requirements U.S. strategic forces should satisfy. These comprised:

> an essential equivalence with the Soviet Union in the basic factors that determine force effectiveness.
>
> . . . a highly survivable force that can be withheld at all times and targeted against the economic base of an opponent so as to deter coercive or desperation attacks on the economic and population targets of the United States and its allies.
>
> . . . a force that, in response to Soviet actions, could implement a variety of limited preplanned options and react rapidly to retargeting orders so as to deter any range of further attacks that a potential enemy might contemplate. This force should have some ability to destroy hard targets, even though we would prefer to see both sides avoid major counterforce capabilities. We do not propose, however, to concede to the Soviets a unilateral advantage in this realm.
>
> . . . a range and magnitude of capabilities such that everyone—friend, foe and domestic audiences alike—will perceive that we are the equal of our strongest competitors.[42]

Aware that these proposals, particularly the third item in the list above, would be controversial, Schlesinger endeavored to separate the question of a change in targeting doctrine from the contentious issue of hard-target counterforce capabilities and from the R&D initiatives in the strategic field he considered prudent. On the desirability of change in targeting doctrine, Schlesinger stressed that changes would be in the direction of flexibility and selectivity and not necessarily, in fact not even preferably, in the direction of greater hard-target counterforce capabilities. Schlesinger maintained that while the manner in which deterrence might fail could not be predicted, it was clear that a massive bolt-out-of-the-blue attack on the United States was only one possibility and probably the least likely one. His main concern was that the Soviet Union now (1974) possessed large and varied strategic capabilities, and that U.S. strategic doctrine had not evolved adequately in response to this new state of affairs.

It was at least conceivable that the Soviet Union would use or threaten to use its strategic forces in ways quite distinct from a massive attack on the United States. If Soviet authorities judged that U.S. strategic forces were postured to respond only to such a massive attack, they might be tempted to use or threaten to use their own forces in more limited ways. Accordingly, Schlesinger proposed that the United States take the steps necessary to insure that it had a variety of strategic options, ranging downward from the massive assured destruction counterstrike to the delivery of just a handful of weapons, with the smaller-scale options directed predominantly at military and economic rather than population targets. In Schlesinger's view, the knowledge that U.S. strategic forces were capable of responding rapidly on any scale and against any type of target would strengthen deterrence against the wider spectrum of threats that the "growth to maturity of Soviet strategic offensive forces" had engendered.[43] A second virtue—and, as shall be seen, the one that seems to have carried the day for the Schlesinger doctrine—was that in the event deterrence were to fail, it might be possible "to bring all but the largest nuclear conflicts to a rapid conclusion before cities are struck."[44]

As already indicated, Schlesinger was adamant that the change in targeting doctrine did not require any new strategic forces, nor in fact any new capabilities for existing forces. All that was involved was acceptance of the need for more strategic options and undertaking the necessary planning—designing alternative packages that varied in the number and type of targets that would be attacked and selecting the weapons to be used—to have these options available for prompt implementation. The only improvements strongly linked with the change in targeting doctrine were "appropriate sensors to assist in determining the nature of the attack, and adequately responsive command-control arrangements."[45] An important element of the latter, the Command Data Buffer System that reduced the time required to retarget a Minuteman III ICBM from 16 to 24 hours to 25 minutes, was in fact under development before Schlesinger proposed his modifications to targeting doctrine.

Schlesinger did admit that a more efficient hard-target-kill capability than the United States then possessed might be desirable "both to threaten specialized sets of targets . . . with a greater economy of force, and to make it clear to a potential enemy that he cannot proceed with impunity to jeopardize our own system of hard targets."[46] But he denied any requirement for or any interest in a major counterforce potential from a unilateral U.S. perspective. Such a requirement, he contended, would emerge only in the event the Soviet Union moved to acquire such a capability. Nevertheless, he considered the momentum and direction of developments in the Soviet strategic forces sufficiently disturbing to warrant some additional R&D effort to improve the accuracy and yield-to-weight ratios of U.S. strategic warheads. It is important to note that Schlesinger was not really concerned that, unless the United States acted, the Soviet Union could acquire a counterforce lead that had genuine military significance (that is, an ability to disarm the United States). No foreseeable technical developments could threaten alert bombers or deployed SSBNs. The risk, in Schlesinger's view, was that if the Soviet Union was permitted to acquire a substantial lead in counterforce capabilities, its leaders might believe that such a favorable asymmetry could be exploited for political advantage. The United States would then face the unnerving task of disabusing the Soviet leaders of this belief in a crisis. As one sympathetic observer put it: "Americans may believe that Soviet leaders so persuaded would be profoundly in error, but the task of convincing them of the truth in this belief could easily entail grappling with a crisis or crises compared with which the events of October 1962 would pale into insignificance."[47] Schlesinger felt it was worth a price in research and development hedges to prevent such illusions arising in the first place.[48]

Although inclined for the most part to play down the revolutionary nature of his proposals concerning strategic targeting, Schlesinger made it abundantly clear they were sharply at variance with the plans he inherited on assuming office. He did acknowledge the fact, documented above, that several response options in addition to assured destruction were added in 1961 during McNamara's brief flirtation with damage limitation, and that these options remained in the SIOP.[49] He also acknowledged that, over the period when assured destruction was the declared strategic doctrine, the employment of worst-case analysis and the testing of force adequacy against the higher-than-expected threat provided a built-in surplus of warheads that were allocated to nonurban targets, including some hard targets.[50] Schlesinger's objection, however, was that all the available options involved large numbers of weapons. If these options were exercised "it would be impossible to ascertain whether the purpose of a strategic strike was limited or not," so in Schlesinger's view they were of no value.[51]

Since the concern here is to assess the impact of these changes in strategic doctrine on the long-range cruise missile programs, a full-scale critique of these changes would be out of place. However, a few brief comments seem in order. Predictably, the principal criticism directed at the Schlesinger doctrine was that it would lower the nuclear threshold—the decision to fire two or three ICBMs would be easier to take. Schlesinger could not, and did not, deny the latter, but he disputed the inference that this would lower the nuclear threshold. His entire thesis rested on the hypothesis that the likelihood of having to face the decision to use strategic weapons on any scale would be reduced if more credible, that is more usable, options were available, and if the opponent knew this to be the case. A closely related line of criticism was that endorsement of Schlesinger's views would make nuclear war seem more respectable. This sentiment was reinforced by the stunningly low estimates provided by Schlesinger of U.S. casualties in the event of a Soviet counterforce strike. Far from the 100 million-odd U.S. deaths traditionally associated with strategic war, Schlesinger claimed that if the Soviet Union attacked U.S. ICBM silos with 2 warheads per silo, and the 46 SAC bomber bases and the 2 SSBN bases with one warhead each, the number of deaths would be only five or six million.[52] If the attack was limited only to ICBMs, and if only one warhead was directed to each silo, U.S. deaths could number just 800,000. In his determination to show that the Soviet Union could use its strategic forces in ways that made a massive assured destruction response quite inappropriate and thus incredible, Schlesinger clearly exaggerated the finesse with which a nuclear exchange could be conducted.

A second line of criticism was that the adoption of war-fighting or counterforce doctrines would create a more permissive environment for the refinement of strategic weapons, and thus fuel the arms competition between the United States and the Soviet Union. This is undoubtedly the case, but it would be easy to overstate the degree to which the arms competition would be stepped up. American technological developments had long been predominantly in the direction of counterforce. While Schlesinger proposed one or two R&D initiatives as hedges against the possibility that improved counterforce capabilities might be desirable in the future, it is worth pointing out that a large number of programs with significant implications for U.S. counterforce capabilities were already under way, for example, Trident, B-1, SLCM, and ALCM. Even the MK-12A higher yield warhead for the Minuteman III was in advanced development before Schlesinger assumed office.

Second, to the extent that this portrayal of Soviet nuclear strategy is accurate, extensive damage-limiting counterforce capabilities were already established as a first priority objective in that country. Perhaps the greatest source of acceleration in the arms competition lay in Schlesinger's insistence that "we do not propose to let an opponent threaten a major component of our forces without our being able to pose a comparable threat."[53] Thus if Soviet strategic doctrine called for the open-ended acquisition of counterforce capability, then the United States would follow suit.

A final comment concerns how Schlesinger's proposals for strengthening the U.S. strategic deterrent shape up against the Soviet approach to deterrence. Apart from enhancing deterrence, Schlesinger's other hope for flexible strategic options was that, if deterrence failed, they offered the prospect of war termination before large-scale attacks were made on cities. Clearly, this hope is founded on Soviet acceptance of the feasibility of limited nuclear war and that the Soviets perceive the virtue of keeping open the possibility of stopping escalation before cities are struck. Unfortunately, there is almost no evidence that the Soviets hold either of these views. Indeed, their reaction to the notion of limited strategic war has been distinctly hostile.[54] Dennis Ross records the reaction of one leading Soviet commentator, G. A. Trofimenko, as follows:

> U.S. attempts to bound and legalise nuclear missile war by "ascribing to the Soviet Union intentions and readiness to wage a 'limited strategic war'" are doomed to fail because in any "test of strength, the Soviet Union will not act in accordance with American 'rules' . . . but in accordance with its own military doctrine—'with the aim of fully smashing any aggressor.'"[55]

Another analyst who has addressed this particular question concludes the Soviets prefer:

> unilateral, as opposed to co-operative, damage-limiting strategies in the event deterrence fails. . . . the preponderance of Soviet thought on this question has shown a preference for the unilateral approach to damage limitation by means of unrestrained counterforce strikes and, where technically feasible, passive and active defenses.[56]

In this light the viability of Schlesinger's notions of mutual restraint and intrawar deterrence is open to serious doubt particularly since, despite occasional hints to the contrary, he endorsed the U.S. tradition of planning only for second strikes. In a sense the Soviet strategic nuclear threshold is higher than that desired for the United

States by Schlesinger. At the strategic level the only strike the Soviet Union contemplates is a massive one. In response to such an attack, the appropriateness or credibility of a U.S. SIOP-level retaliation is not seriously questioned, not even by Schlesinger.

AFTER SCHLESINGER

James Schlesinger abruptly departed the scene in November 1974, but by this time the work on his principal doctrinal innovation—the assembly of a number of selective strategic options—was well under way. His successor, Donald Rumsfeld, fully endorsed his rationales for the need for selectivity and flexibility in the strategic forces, including a limited capability to strike hard targets. Accordingly, Rumsfeld persevered with the strategic programs he inherited from Schlesinger. In his second annual report, that for FY1978 and released in January 1977, Rumsfeld expressed something approaching alarm because the Soviet Union appeared to be determined to acquire a comprehensive hard-target kill capability.[57] Before the mid-1980s a hypothetical Soviet attack on U.S. ICBMs could reduce the number of surviving missiles to a low level, while the Soviet Union retained the greater part of its ICBM force.[58] Like Schlesinger, and indeed for almost exactly the same reasons, Rumsfeld regarded such a prospect as unacceptable even though he acknowledged that the ability of the United States to launch a retaliatory assured destruction strike could not be jeopardized. He warned the Soviets that trends in their strategic capabilities could drive the United States to the deployment of mobile ICBMs. Moreover, if this turned out to be necessary, the United States would see to it that the Soviet Union would also have to bear the heavy cost of changing the basing mode for ICBMs: "Since high accuracies can be built into mobile as well as fixed systems, the Soviet leadership should be aware that if the United States moves toward mobility, the Soviets will have strong incentives to go mobile as well."[59]

To support the position that the United States would not tolerate the emergence of a large asymmetry in favor of the Soviet Union in ICBM counterforce capability, Rumsfeld proposed to accelerate the development of MX, a new ICBM with a much higher throw-weight than Minuteman III that would be deployed either in underground trenches or hardened shelters. Specifically, he requested $245 million to initiate engineering development of this weapon, with the aim of achieving an initial operating capability in December 1983.[60]

The Carter administration introduced notably different judgments and perceptions on several of the major issues concerning the strategic forces. The strategic war plans (or the SIOP) inherited from the previous administration were detailed in National Security Defense Memorandum 242 (NSDM-242), which was drawn up in 1974 under Schlesinger. Apart from increasing the number of genuinely limited options, NSDM-242 also reflected a new twist on the capabilities required for assured destruction. The notion that assured destruction should be measured on a relative scale, first raised in 1971 by Melvin Laird, had by 1974 crystallized into the requirement that the Soviet Union should not recover from nuclear war more rapidly than the United States. Thus NSDM-242 called for the destruction of 70 percent of Soviet industry needed for postwar recovery.[61]

In its initial review of these plans—as part of the PRM-10 study mentioned above—the Carter administration endorsed this objective.[62] Subsequently, in December 1977 (PRM-10 was completed in June 1977) the detailed targeting plans in NSDM-242 were the subject of a major review. Senior officials in the administration, notably Zbigniew Brzezinski, the president's national security adviser, felt the targeting priorities had been assigned without regard to political and psychological considerations, and that the deterrent effect of the U.S. strategic threat could be enhanced if this were done.[63]

As regards the notion of limited strategic warfare, neither President Carter nor Secretary of Defense Brown found it attractive or feasible. At his confirmation hearings before the Senate Armed Services Committee in January 1977, Brown stated that "I do not think it at all likely that a limited strategic nuclear exchange would remain limited."[64] However, this conviction was not held sufficiently strongly to lead to a recommendation for a return to an assured destruction minimum deterrent. One month later, in his first appearance before Congress as secretary of defense, Brown argued that:

> While it will likely be very difficult to restrain any nuclear conflict at a level of limited exchanges, that is not a reason to preclude planning for situations short of full scale nuclear war. It would be counterproductive to leave only two responses to nuclear attack: massive response or no response. Planning which provides for less than full scale war thus can enhance deterrence by raising confidence in our capability to respond across a continuum of possible scenarios.[65]

The Defense Department report for FY1979 provided Brown with his first opportunity to spell out the requirements for a credible strategic nuclear deterrent. In this report he made clear his view that no matter how small the chances were of limiting strategic war, the enormity of the consequences of unlimited strategic war demanded that this possibility be kept alive:

> [Assured destruction] must not be automatic, our only choice, or independent of an enemy's attack. . . . we are quite uncertain as to how an adversary with increasingly sophisticated strategic forces might consider employing them in the event of a deep and desperate crisis. But we know that a number of possibilities would be open to him. As a consequence, we must have the flexibility to respond at a level appropriate to the type and scale of his attack. . . . Though the probability of escalation to a full-scale thermonuclear exchange would be high in these circumstances we must avoid making that probability a certainty.[66]

While one could highlight the differences between Schlesinger (or Rumsfeld) and Brown on the utility of limited strategic warfare, it seems apparent that, in practical terms, there was no diminution in the importance attached to flexibility and selectivity in the employment of strategic weapons. In the present context, a more important change was the noticeably more relaxed attitude on the part of Secretary Brown on the question of the prospective asymmetry between the United States and the Soviet Union in the counterforce capability of their respective ICBM forces. This was given immediate expression in the amendments proposed to the FY1978 defense budget inherited from the Ford administration. Specifically, Brown recommended that engineering development of the MX be postponed until the most desirable basing mode had been determined. In addition he recommended closing down the Minuteman III production line, thus ruling out the possibility of expanding the deployment of this weapon beyond the 550 planned.

One reason for this move was that, in all probability, the decision had already been taken to try once more to address the issue of the emerging vulnerability of silo-based ICBMs in the context of SALT II. Among the SALT proposals Secretary of State Vance submitted to the Soviet Union in March 1977 was one that sharply limited the permissible number of MIRVed ICBMs and, in addition, provided for constraints on their modification, replacement, and flight testing, all of which was designed primarily to preserve the viability of the fixed, land-based ICBM. This proposal was rejected, but Brown nevertheless declined to put the MX back on an accelerated development schedule. As was seen in the preceding chapter, Brown was as persuaded as anyone of the political costs and potential risks associated with the Soviet Union having an unmatched capability against a key element of the U.S. strategic Triad. But, on the one hand, he did not see this as dangerously imminent, and on the other hand, he was "not persuaded that the right way to deal with a major Soviet damage-limiting program would be by imitating it."[67]

In consequence, Brown sought to counter the quantitative and qualitative growth in Soviet strategic forces in ways that would minimize Soviet fears that the United States was acquiring a disarming first-strike capability. In this context the long-range cruise missile, launched from strategic bombers, came into its own. The rationales for the major role assigned to the air-launched cruise missile were discussed in the previous chapter, but it is necessary to briefly recall them here. For Brown, equipping 150-odd B-52s with 20 ALCMs each would preserve both the appearance and the substance of strategic parity with the Soviet Union by adding some 3,000 warheads to the U.S. arsenal with an accuracy/yield combination sufficiently potent to handle the hardest target, and with sufficient range to reach 70 to 80 percent of all targets in the SIOP. In addition, a cruise missile force could be expanded relatively quickly if the necessity arose, it would be a visible indication of U.S. technological superiority, and it did not threaten a first-strike capability.

Within a year, however, the Carter administration's relatively sanguine attitude toward the high probability of a major Soviet lead in first-strike counterforce capability by the mid-1980s was replaced by the attitude that something more had to be done. In his annual report for FY1980, in the course of a candid discussion on the great difficulty of devising a sensible and credible targeting policy for strategic nuclear weapons, Defense Secretary Brown made the following points:

> We must insist on essential equivalence with the Soviet Union to symbolize the equality that both sides accept in this realm. But we must not mistake the symbols, however important, for the substance. We may be able to obtain deterrence, and achieve assured destruction or more, without equivalence; it is by no means certain that equivalence alone will give us deterrence.
>
> There is no obvious solution to our dilemma at this juncture. . . . One resolution of this issue . . . would be, first, being able to cover hard targets with at least one reliable warhead with substantial capability to destroy the target and, second, in having the retargeting capability necessary to permit reallocation of these warheads either to a smaller number of crucial hard targets, or to other targets on the list. Even with slow-reacting capabilities such as cruise missiles, this would ensure that an enemy's silos are not a kind of sanctuary from which he can shoot with impunity. . . . the times and the uncertainties surrounding nuclear deterrence warrant such an approach.[68]

While the cruise missile was not downgraded in any way, it is apparent Secretary Brown no longer considered it to be enough or, more accurately, to be a sufficiently direct response to the Soviet effort to secure a comprehensive first-strike capacity against hard targets in the United States.

This was primarily a change in declaratory policy—the United States had targeted increasing numbers of Soviet ICBMs since 1962 and probably every Soviet ICBM since the mid-1970s at the latest. That is, while the United States was probably already capable of denying the Soviet Union any advantage from striking first, it was judged to be necessary also to declare that the United States would deny the Soviet Union such an advantage by proclaiming a targeting doctrine that implied that Soviet silos would no more be sanctuaries than U.S. silos.[69]

Brown's proposed revisions in targeting doctrine were, in effect, the fulfillment of Schlesinger's 1974 declaration that "we do not propose to let an opponent threaten a major component of our forces without our being able to pose a comparable threat."[70] In the intervening four years the Soviet Union had demonstrated—by insisting on a relatively large number of MIRVed ICBMs in SALT II and by working very hard and very effectively to increase the accuracy of reentry vehicles and the number per missile—that it was determined to acquire the ability to pose a credible threat to U.S. fixed, land-based ICBMs, inport SSBNs, and nonalert bombers. Endorsement of the new targeting policy would permit the Pentagon to acquire more counterforce capability openly and with deliberate speed instead of by the (relatively) subdued process of normal technological improvements.

ASSESSMENT

With this background it should be possible to gauge the impact of the move away from an assured destruction minimum deterrent on the support for and the roles assigned to long-range cruise missiles. On the whole the influence appears not to have been decisive; long-range cruise missiles did not acquire a secure rationale from the move to flexible response and the notion of limited strategic warfare. This, however, is a rather bald statement and there are significant nuances that are worth exploring.

Over the period from 1969 to 1973, when the Nixon administration was making some tentative moves in the direction of a more flexible strategic deterrent, the leading champion of the cruise missile was DDR&E John Foster. Foster's enthusiasm, however, was based more or less exclusively on the cruise missile's ability to contribute to the certainty of U.S. strategic retaliation. It will be recalled that he was persistently concerned with the fact that all existing strategic forces in the U.S. arsenal involved putting weapons high up in the atmosphere, thus making them potentially vulnerable to sophisticated air defenses. Not only was a low-flying weapon the most difficult of all to counter, in Foster's view, but the addition of such a weapon to the strategic arsenal would enhance the ability of all the others to penetrate Soviet defenses.

With the arrival of Schlesinger and his denunciation of assured destruction in favor of strategic forces capable of measured responses to a wide range of threats, one might have expected a relatively enthusiastic endorsement of the cruise missile program. After all, the TERCOM guidance system was tested in March 1973 and gave a clear indication of the remarkable accuracy that could be achieved. Essentially, however, Schlesinger was not interested. Although he argued that, in principle, one could conceive of a nuclear attack beginning with the delivery of two or three weapons and that the United States should be able to respond on a similar scale, it was readily apparent that his only real concern was the number and throw-weight of Soviet ICBMs. When married with MIRV and improved accuracy, the Soviet Union could, at least in principle, strike first and create such an imbalance in residual forces that the United States could threaten no credible or rational response. In other words, Schlesinger's preeminent concern was with prompt and survivable counterforce capabilities; and cruise missiles did not fit this bill. In addition, Schlesinger did not want to jeopardize his proposals on targeting by associating them with new strategic weapons, particularly with a Congress that was not, at the time, sympathetically inclined toward the Pentagon.

Thus, in his FY1975 report, Schlesinger was content to describe the arrangements that had been made to develop the cruise missiles without offering much in the way of rationales. Similiarly his DDR&E, Malcolm Currie, while being more forthcoming on the virtues of the cruise missile, made no attempt to link them with the issues of flexibility and selective targeting. In contrast, in connection with the FY1976 defense budget, both Schlesinger and Currie referred to the SLCM's "unique potential for unambiguous, controlled, single weapon response."[71] This rationale was obviously an outgrowth of the Schlesinger doctrine, but it did not survive very long and it was not wielded with any great conviction. In addition to the reasons already given, it should be recalled that by this time (early 1975) an open division existed between the Air Force and elements within the Office of the Secretary of Defense on the survivability of the cruise missile. If the ability of the cruise missile to penetrate to its target, even when launched in some numbers, was questionable, what would be the fate of a single weapon directed against the most comprehensive air defense system in the world on full alert? The last thing one wants to do in a major crisis is inspire the opponent's confidence; an unsuccessful selective strike may do just that.

When the Carter administration elected to give the cruise missile a major role in maintaining the strategic deterrent, it opted for the ALCM rather than the SLCM for reasons that had nothing to do with limited nuclear options. On the contrary, it was the quantity of ALCMs that could be loaded aboard a part of the strategic bomber force that seemed to be decisive. In addition, of course, the weapon's qualitative characteristics, particularly accuracy, were very important in that Defense Secretary Brown considered it could play a major role in restoring the balance of forces after a Soviet first strike (that is, after a counterforce exchange). The one strong link that can be identified among Schlesinger, Brown, and the cruise missile is in the area of perceptions. Schlesinger attached major, indeed unprecedented, importance to the fact that all parties—the United States, the Soviet Union, and other nations—should perceive a balance in the strategic forces of the two superpowers. In his second report he made the achievement of perceived equality an essential requirement of deterrence. This was justified as follows:

> equality is also important for symbolic purposes, in large part because the strategic offensive forces have come to be seen by many—however regrettably—as important to the status and stature of a major power . . . the lack of equality can become a source of serious diplomatic and military miscalculation. Opponents may feel that they can exploit a favorable imbalance by means of political pressure.[72]

Brown attached no less importance to perceptions. He regarded the decision to acquire the B-52/ALCM force as making perhaps a greater contribution in this regard than any other of the strategic programs under way or contemplated.[73]

A secondary link between strategic doctrine and the emergence of the cruise missile as a major new component of the strategic forces might have been the emphasis given to maximizing the time it would take for the Soviet Union to recover from nuclear war. In the past the destructive capabilities required tended to be defined in terms of the extent of the damage to be inflicted. Retarding the rate of recovery implied that the degree of damage was equally important. This new element in U.S. nuclear targeting philosophy would not have affected the number or type of targets in the SIOP. On the other hand, it would have supported a requirement for larger and (particularly) more accurate second-strike assured destruction weapons to insure that facilities critical to a recovery effort were destroyed beyond repair. The cruise missile, being extremely accurate and carrying a warhead six times larger than a Poseidon SLBM reentry vehicle—the mainstay of the retaliatory forces—fitted this requirement admirably.

To sum up, the strategic cruise missile did not acquire a strong doctrine-related rationale until mid-1977 when it displaced the B-1. At that time the view was put forward that the cruise missile would effectively counter the emerging Soviet preponderance in hard-target kill capability. In the preceding five years, with one short-lived exception, it had never been suggested that the cruise missile would fill a void in U.S. strategic capabilities opened up by revisions in strategic doctrine. The exception was the SLCM's "unique potential for unambiguous, controlled single-weapon response." Finally, although it can be argued that the cruise missile is well suited to the emphasis given since 1974 to retarding the rate of Soviet economic recovery from a U.S. countervalue strike, no such link was ever officially suggested.

NOTES

1. A notable difference is that, whereas in the United States weapons technology has tended to outstrip the requirements of declared strategic doctrine leading to revisions of the latter, Soviet strategic doctrine seems to have remained constant and the preeminent concern has been to secure nuclear force capabilities consistent with this doctrine.

2. Quoted in Jerome H. Kahan, Security in the Nuclear Age: Developing U.S. Strategic Arms Policy (Washington, D.C.: Brookings Institution, 1975), p. 78.

3. Status of U.S. Strategic Power, hearings, Preparedness Investigating Subcommittee of the Senate Armed Services Committee, April 1968, p. 7.

4. Ibid., p. 6. Alain Enthoven, McNamara's chief systems analyst, has also acknowledged that damage limitation was an active component of targeting policy over the years of assured destruction. See Alain C. Enthoven and K. Wayne Smith, How Much is Enough? (New York: Harper & Row, 1971), p. 195.

5. Hearings on Military Posture and An Act (S. 3293), House Armed Services Committee, April-June 1968, p. 9,581.

6. Robert S. McNamara, The Essence of Security: Reflections in Office (London: Hodder and Stoughton, 1968), pp. 51-59. For a succinct review of the renewed emphasis on superiority at about this time, see Kahan, Security in the Nuclear Age, pp. 106-9.

7. This review became National Security Study Memorandum-3 (NSSM-3). An indication of its content as far as strategic forces are concerned is available in the New York Times, 1 January, 1 May, and 2 May 1969.

8. U.S. Foreign Policy for the 1970s: A New Strategy for Peace, A Report to the Congress by Richard Nixon, president of the United States, 18 February 1970, p. 122.

9. The term "sufficiency" was apparently first used in the strategic context in 1956 by Secretary of the Air Force Donald Quarles. It was accepted by the Nixon administration in the first half of 1969 primarily, it seems, to put to rest any notions that superiority would be written into declared U.S. strategic policy. The first public indication of what sufficiency meant in terms of guidelines for force posturing (items [a] through [d] in the text) came in the defense secretary's annual report for FY1972.

10. Melvin R. Laird, secretary of defense, Toward a National Security Strategy of Realistic Deterrence: Fiscal Year 1972-1976 Defense Program and the 1972 Defense Budget, 9 March 1971, p. 62.

11. Below the level of public declarations there was already a quite definite commitment to flexible options, as the following exchange in the Senate early in 1971 makes clear:

> Senator Symington. Is it our policy, however, to develop warheads with a hard target kill capability for selective use, that is, in the context of a limited nuclear war?
>
> John Foster Jr. (Director of Defense Research and Engineering). Yes, sir. . . .

Fiscal Year 1972 Authorization for Military Procurement, hearings, Senate Armed Services Committee, March-May 1971, p. 465.

12. P. Ramsey, "A Political Ethics Context for Strategic Thinking," in Strategic Thinking and Its Moral Implications, ed. M. Kaplan (Chicago, 1973), p. 135.

13. A notable example of this school was Fred Charles Iklé, director of the U.S. Arms Control and Disarmament Agency from 1973 to 1977. See his article, "Can Nuclear Deterrence Last Out the Century?," Foreign Affairs (January 1973): 267-85.

14. Wolfgang K. H. Panofsky, "The Mutual Hostage Relationship between America and Russia," Foreign Affairs (October 1973): 109-18.

15. Henry Kissinger, Nuclear Weapons and Foreign Policy (New York: Harper & Brothers, 1957), and The Necessity for Choice (New York: Doubleday, 1960).

16. James R. Schlesinger, Arms Interaction and Arms Control (Santa Monica, Calif.: Rand Corporation, P-3881, September 1968).

17. Col. Gen. N. A. Lomov, ed., Scientific-Technical Progress and the Revolution in Military Affairs, Moscow, 1973 (trans. and published under the auspices of the United States Air Force), p. 272.
18. This quotation, taken from Communist of the Armed Forces, November 1975, is cited by Foy D. Kohler in the foreword to Leon Goure, War Survival in Soviet Strategy (Florida: Coral Gables, 1976).
19. Lomov, Scientific-Technical Progress and the Revolution in Military Affairs, p. 276.
20. Ibid., p. 274.
21. Ibid., p. 269.
22. Stanley Sienkiewicz, "SALT and Soviet Nuclear Doctrine," International Security (Spring 1978): 95.
23. Lomov, Scientific-Technical Progress and the Revolution in Military Affairs, p. 271.
24. Dennis Ross, "Rethinking Soviet Strategic Policy: Inputs and Implications," Journal of Strategic Studies (May 1978): 6. Ross suggests the terms "deterrence through denial" and "deterrence through punishment" to characterize, respectively, the Soviet and U.S. positions (p. 9).
25. Leon Gouré, Soviet Civil Defence 1969-70 (Florida: Coral Gables, 1971). Gouré's first volume on the subject appeared in the early 1960s: Civil Defense in the Soviet Union (California: University of California Press, 1962).
26. Lomov, Scientific-Technical Progress and the Revolution in Military Affairs, pp. 273-74.
27. Thomas K. Jones et al., Industrial Survival and Recovery after Nuclear Attack: A Report to the Joint Committee on Defense Production, U.S. Congress (D180-20236-1), 1976.
28. Ibid., p. 73.
29. Leon Gouré, War Survival in Soviet Strategy. Another useful, though somewhat cryptic, review of city defense in the United States and the USSR that appeared at this time is United States and Soviet City Defense: Considerations for Congress, prepared by Congressional Research Service, Library of Congress, 30 September 1976.
30. Bulletin of Atomic Scientists, April 1978, p. 51.
31. Richard Pipes, "Why the Soviet Union Thinks it Could Fight and Win a Nuclear War," Commentary, July 1977, pp. 21-34. Pipes was chairman of the so-called Team B intelligence group. In 1976 President Ford took the unusual step of authorizing two groups to review Soviet strategic capabilities and intentions, with both groups to be provided with the same intelligence data. Team A was made up of officials from the relevant government departments and agencies, while Team B consisted of knowledgeable outsiders. Other than Pipes, Team B. consisted of Thomas W. Wolfe, Lt.-Gen. Daniel O. Graham,

Paul D. Wolfowitz, Paul H. Nitze, Gen. John Vogt, and Professor William Van Cleave. Team B's assessment was dramatically more pessimistic than that of Team A and the compromise, reflected in the National Intelligence Estimates released in December 1976, was officially described as the most somber evaluation of Soviet capabilities and intentions in more than a decade.

32. This quotation is taken from Pipes's response to the numerous letters to the editor that his July article provoked. See Commentary, September 1977, p. 22.

33. Ibid., p. 24.

34. Cited in International Herald Tribune, 7-8 January 1978, pp. 1, 3.

35. International Herald Tribune, 20 July 1978, p. 1.

36. Interview, U.S. News and World Report, 5 September 1977, p. 21.

37. Harold Brown, secretary of defense, Department of Defense Annual Report Fiscal Year 1979, 2 February 1978, p. 64. Also see p. 48, where the main features of the Soviet civil defense program are described.

38. International Herald Tribune, 21 June 1978, p. 1.

39. Schlesinger expounded on the modifications of U.S. strategic doctrine he desired in a number of documents and forums. The more important of these were the following: Annual Defense Department Report, FY1975, 4 March 1974; Annual Defense Department Report, FY1976 and FY1977, 5 February 1975; U.S.-USSR Strategic Policies, hearings, Subcommittee on Arms Control, International Law and Organizations of the Committee on Foreign Relations, United States Senate, 4 March 1974; and Briefing on Counterforce Attacks, hearings, Subcommittee on Arms Control, International Law and Organizations of the Committee on Foreign Relations, United States Senate, 11 September 1974.

40. Annual Defense Department Report, FY1975, p. 27.

41. Ibid., p. 28.

42. Annual Defense Department Report, FY1976 and FY1977, pp. I-13-14.

43. Annual Defense Department Report, FY1975, p. 26.

44. Ibid., p. 38.

45. Ibid.

46. Ibid., pp. 40-41.

47. Colin S. Gray, "Nuclear Strategy: The Debate Moves On," Royal United Services Institute Journal of Defence Studies (March 1976): 49.

48. Annual Defense Department Report, FY1975, p. 43.

49. The fact that the SIOP remained essentially unchanged from 1962 through the early 1970s has been documented by Desmond Ball.

See his "Deja Vu: The Return to Counterforce in the Nixon Administration," in The Strategic Nuclear Balance: An Australian Perspective, ed. Robert O'Neill (Canberra: ANU Press, 1975), especially pp. 171-73.

50. Annual Defense Department Report, FY1975, pp. 35-36.

51. U.S.-USSR Strategic Policies, hearings. The salient portions of Schlesinger's testimony are reproduced in Robert J. Pranger and Roger P. Labrie, eds., Nuclear Strategy and National Security: Points of View (Washington, D.C.: American Enterprise Institute for Public Policy Research, 1977), pp. 104-26 (the quote appears on p. 106). Although the evidence is inconclusive, it does seem that Schlesinger exaggerated the massiveness and inflexibility of the strategic targeting plans he inherited in 1973. See Ball, "Deja Vu," pp. 172-78.

52. Briefing on Counterforce Attacks, hearings, pp. 11-13. The casualty estimates included deaths from both the direct effects of nuclear explosions and from fallout. At the time the Soviet potential to attempt strikes such as these was considered to be some seven to eight years away.

53. Annual Defense Department Report, FY1975, p. 42.

54. On the other hand, the Soviet reaction to the U.S. announcement that it planned to move to counterforce targeting was partly one of surprise. As Schlesinger remarked (Briefing on Counterforce Attacks, hearings, p. 17), the Soviets pointed out that they always assumed the United States would target in this way. This, in turn, suggests the obvious point that the Soviet Union bases its assessment on U.S. capabilities and their probable implications for action policy rather than on what the United States declared its strategic policy to be. Interestingly enough, in his 1968 Rand paper cited earlier, Schlesinger had opined that the Soviet Union would ascribe counterforce targeting to the United States whether or not this was in fact done.

55. Ross, "Rethinking Soviet Strategic Policy," p. 22.

56. Jack L. Snyder, The Soviet Strategic Culture: Implications for Limited Nuclear Options (Santa Monica, Calif.: Rand Corporation, R-2154-AF, September 1977), p. 38.

57. Donald H. Rumsfeld, secretary of defense, Annual Defense Department Report, FY1978, 18 January 1977, p. 70. This report also acknowledged, more explicitly than any other recent Defense Department annual report, that Soviet strategic doctrine was very different from that of the United States, and that no convergence to the latter's perspective could be anticipated (p. 59).

58. Ibid.

59. Ibid., p. 72.

60. Ibid., p. 130.

61. Los Angeles Times, 2 February 1977, p. 1.
62. International Herald Tribune, 7-8 January 1978, p. 3.
63. New York Times, 16 December 1977, p. 5. For example, given the manifest inefficiency of Soviet agriculture and the grave political risks associated with food shortages, the Soviet food storage and distribution system should be assigned a higher priority than other economic targets.
64. Ibid.
65. Department of Defense Appropriations for 1978, hearings, House Committee on Appropriations, February 1978, p. 108.
66. Department of Defense Annual Report, Fiscal Year 1979, pp. 55-56.
67. Ibid., p. 65.
68. Harold Brown, secretary of defense, Department of Defense Annual Report, Fiscal Year 1980, 25 January 1979, pp. 76-78.
69. Eighteen months later, in July 1980, President Carter signed Presidential Directive 59 and committed the United States to a nuclear targeting strategy that gives priority to striking military and political targets in the Soviet Union.
70. Pranger and Labrie, eds., Nuclear Strategy and National Security: Points of View, p. 101. Even Kissinger had come to the view that a major imbalance in prompt counterforce capabilities could have serious geopolitical consequences ("Kissinger Critique," Economist, 3 February 1979, p. 20).
71. Annual Defense Department Report, FY1976 and FY1977, p. II-39.
72. Ibid., p. II-7.
73. Department of Defense Annual Report, Fiscal Year 1979, pp. 56-57, 115-16.

7

TECHNOLOGICAL MOMENTUM

The current-generation cruise missiles are characterized by small overall volume, extremely compact turbine propulsion systems, and compact, high-capability electronic systems primarily related to guidance. This chapter will briefly examine the extent to which early cruise missiles evolved in these directions and the relevant technological developments that occurred in the decade or so that separated the early development programs from the current ones. Finally, an attempt will be made to assess the extent to which the relevant technologies were ripe when the current programs were initiated.

As regards the first criterion, that of overall size, there was no evident tendency for the early weapons to become smaller. This is apparent from Table 7.1. Indeed, there is no evidence that demanding constraints on size were placed on any of the weapons when development was initiated. This is particularly evident in the case of Regulus. The only vessel purpose-built to carry Regulus II, the nuclear-powered submarine Halibut, could only take two of the missiles, whereas it was deployed with four Regulus Is when the former was canceled.[1]

A major factor determining the dimensions of a missile is its propulsion system. All the cruise missiles in Table 7.1 were built around existing aircraft engines (Table 7.2). There was no separate technological effort to provide turbine engines for missile applications.[2] Moreover, the primary performance requirement for jet engines in the late 1940s and early 1950s was higher thrust and the ability to produce supersonic performance.[3] Available engines tended to become larger. Over time, as aircraft roles became more specialized, engine developments were similarly refined to produce power plants optimized for the particular mission the aircraft was intended to perform.[4] This evolving sophistication or flexibility in securing desirable advances in turbine engines suggests that power plants more

TABLE 7.1

Early Cruise Missiles: Weights and Dimensions

Missile	Year Operational	Length (meters)	Diameter (meters)	Wing Span (meters)	Launch Weight (kg.)
TM-61 A/C Matador	1954	12.1	1.4	8.7	6,273
TM-76 A/B Mace	1959	13.5	1.4	6.9	8,182
SSM-8a Regulus I	1955	10.4	1.4	6.4	5,455
ZSSM-9 Regulus II	—[a]	18.1	1.3[b]	6.1	10,000
AGM-28 A/B Hound Dog	1959	12.1	0.7[c]	3.7	4,545

[a] Canceled December 1958 when in production.
[b] Excludes large air-intake under the fuselage.
[c] Excludes entire propulsion system, which was carried under the fuselage.

suited to cruise missile application could have been developed if there had been a demand for them.

It is not being suggested that ongoing developments in turbine engines were entirely irrelevant to cruise missile applications. Virtually all sought after and progressively achieved developments—in thrust-to-weight ratios, inlet temperatures, compression ratios, specific fuel consumption, and so on—were of the utmost relevance. But without the focus of an ongoing interest in cruise missiles, highly suitable engines tended not to emerge. Taking guidance from the current cruise missiles such engines would feature, relative to prevailing engine technology, higher thrust-to-weight ratios, lower specific fuel consumption, compact size to provide a missile suited to its intended launch platform, and low cost because it would be used only once. As it happened, several small engines with high thrust-to-weight ratios were developed in the late 1950s for unmanned applications but, as has been seen, interest in the cruise missile had waned considerably by this time.

TABLE 7.2

Cruise Missile Turbine Engines

Missile	Engine Designation	Aircraft Application
Matador	J-33-A-37	F-102, F-106
Mace	J-33-A-41	F-102, F-106
Regulus I	J-33-A-14	F-102, F-106
Regulus II	J-79	F-4, F-104
Hound Dog	J-52	A-4, A-6

One of these applications, the ADM-20 B/C Quail decoy missile, is of interest to this study since the ALCM is linked directly to the subsonic cruise armed decoy the Air Force intended as a replacement for the Quail. The Quail was developed in the late 1950s as a penetration aid for the B-52. After launch from the B-52, the Quail would go its own way with the electronic equipment enhancing its image to make it resemble a B-52 on enemy air defense radar screens. The Quail's engine, the J-85-7 turbojet, weighed only 326.5 lbs., provided 2,450 lbs. of thrust, and at 17.5 inches in diameter and 39.3 inches in length, was extremely compact. The J-85-7's thrust-to-weight ratio of 7.5:1 was well in excess of most other engines of that period.[5] On the other hand, the engine was relatively inefficient. Given the general size constraint of internal carriage in the B-52 (four per aircraft) and the required payload of electronic equipment, the available fuel gave the Quail a range of just 200 n.m. at high altitude. At low altitude, increasingly the preferred mode of penetration, this shrank to 95 n.m.[6] Production of the Quail ended in February 1961.

For the next several years, until about 1969, there were no identifiable cruise missile programs in the United States. As far as engine developments are concerned, one partial surrogate for cruise missiles must be mentioned: the remotely piloted vehicle (RPV). United States interest in RPVs received two major boosts during the 1960s. The first was the embarrassment caused when a manned U-2 intelligence aircraft was brought down over the Soviet Union in May 1960. The second was the war in Indochina. The major applications for RPVs have been in reconnaissance and electronic intelligence, although considerable interest has been displayed in recent years in their use as weapon delivery platforms.

RPVs and cruise missiles have several common characteristics. The RPV is not expected to survive a large number of flights since it will be used primarily in the most heavily defended environments. This fact places a high premium on low cost. Equipment, including the engine, must be highly reliable, but not for much longer than the expected life span of the vehicle. Similarly, it is advantageous if the RPV is compact since air launching, mostly from the specially configured C-130 Hercules, has long been standard. It must be pointed out that this constraint has so far served only to limit growth in size. The Teledyne Ryan Model 147, perhaps the most widely used RPV, still bears an unmistakable likeness to the Model 124 pilotless target drone from which it was derived. Other relevant improvements in RPVs included the reduction of radar cross section and infrared signature. The Continental Aero Engines Division of Teledyne Ryan subsequently emerged as a leading competitor for the engine contract for ALCM and SLCM.

On the other hand, the general tenor of this discussion, that of a lack of official interest in very small gas turbine engines, is reflected in the experience of the Williams Research Corporation, the other leading engine contractor for ALCM and SLCM. Williams had been engaged in the development of such engines for more than 20 years before the military displayed any significant interest in its products in the form of a $1.4 million contract in 1970 to further develop the WR19 turbofan for potential use in SCAD.

The second area of technology central to the current cruise missiles is electronics. Progress in this area has been astonishingly rapid, particularly since the introduction of integrated circuitry in 1961. One expert has summarized this progress as follows:

> Complexity of integrated circuits has approximately doubled every year since their introduction. Cost per function has decreased several thousandfold, while system performance and reliability have been improved dramatically.[7]

These developments have been vigorously pursued for all military applications. Clearly, the lack of interest in cruise missiles in no way retarded the evolution of the "black boxes" installed in the current weapons. Of particular note here is the compact, lightweight computer that is the heart of the TERCOM guidance system.

TERCOM itself was not a new concept when the current cruise missile programs were launched. However, it is more recent than the ATRAN system in the MACE A to which, at least to a layman, it bears a superficial resemblance.[8] In August 1958 the Air Force contracted the Chance Vought Aircraft company to initiate the development of a long-range, low altitude surface-to-surface missile powered

by a nuclear ramjet. This weapon was code-named Supersonic Low
Altitude Missile (SLAM), and was probably the dark horse referred
to in the quotation from Missiles and Rockets in Chapter 2. Very
little information is available in the public record, but one source
refers to an "advanced self-contained guidance sub-system capable of
directing missiles (including SLAM) and aircraft to their targets with
unprecedented accuracy."[9] This guidance system, it appears, was
based on the TERCOM principle.[10] TERCOM was first tested on
board an aircraft in 1961. When SLAM was canceled toward the end
of 1963, a low-level, company-funded development program for
TERCOM was maintained although, apart from the short-lived effort
in connection with Hound Dog, no application for the system emerged
until the early 1970s.

A subsidiary question here concerns the availability of high
resolution topographical maps of targets and approach routes for
cruise missiles utilizing the TERCOM guidance principle. There is
little doubt that information of this kind has been collected for a long
period prior to the current cruise missile programs. Such information
would appear to be essential for the purpose of identifying and accurately determining the position of potential targets from aerial and
satellite reconnaissance photographs. Similarly, such information
would be extremely useful to enable bombers penetrating at low altitude to exploit terrain-related limitations on enemy SAM defenses.

In sum, the availability of suitable topographical maps of enemy
territory probably cannot be regarded as a factor delaying the emergence of TERCOM-guided cruise missiles. However, frequent references to the Defense Mapping Agency in the ALCM/SLCM context
suggest that the data had to be processed in new ways and considerably refined to optimize its utility for TERCOM application.

Not surprisingly, this discussion yields ambivalent conclusions
on the ripeness of cruise missile technologies at the time the current
programs were launched. Strictly speaking, there are not technologies
in the current cruise missiles that could not have been brought to
fruition many years earlier had the ambition existed to do this. As it
happened, suitable turbofan engines emerged indirectly but still required considerable refinement to provide the performance specified,
first for the SCAD and subsequently for the SLCM and ALCM. TERCOM
limped along without the focus of possible application in an operational
weapons system. A statement in 1968 by General Wheeler, chairman
of the Joint Chiefs of Staff, assuming that it is based on the experience
of attempting to fit Hound Dog with TERCOM guidance, is probably an
indication of the system's immaturity at that time. Commenting on the
advanced manned strategic aircraft, later to become the B-1, General
Wheeler said:

> This can be both a bomber . . . but it can also carry
> nuclear armed missiles with which to deliver a standoff
> attack. . . . we are talking, however, about greater accu-
> racies than we have been able to achieve up to the present
> time.[11]

Four years later DDR&E John Foster opined that SLCM (and, by implication, SCAD/ALCM) was a low technical risk program.[12] As has been seen, some allowance must be made for the fact that these two comments were made by the chairman of the Joint Chiefs and the DDR&E respectively, but one might conclude this brief assessment as follows. Over the period from about 1968, when U.S. military interest in cruise missiles was first rekindled, to 1972 when the first strategic cruise missile program was proposed, the development of compact, long-range, low altitude, highly accurate cruise missiles transitioned from a theoretical possibility to a predictable accomplishment. That is to say, toward the end of this period, expert advice was available to predict with high confidence the further developments necessary to make an effective cruise missile a reality.

Restated from a different perspective, the strategic missile in 1972 was a weapon whose time had come. But it was not a weapon that, over the preceding decade, had been periodically considered as a strategic option and deferred because the relevant technologies were not available or seemed to be high risk propositions.

After this, however, it was a very different story. Once the surprising decision had been taken to develop strategic cruise missiles, it quickly became apparent that a supremely elegant weapons system was within reach of a normal development program. It seems most officials were surprised by the potential capabilities of the cruise missile as these emerged from the first two years or so of development work. In 1972 even John Foster, the cruise missile's most fervent supporter, regarded the limited range then considered possible as a serious weakness. It soon became apparent that the minimum-range requirement (1,400 n.m.) could be handsomely exceeded even by the torpedo-tube version, let alone the large diameter models that were under consideration in the earlier stages. The adequacy of the torpedo-tube version also meant significant dividends in terms of reduced radar cross section. The practical feasibility of the TERCOM guidance system was demonstrated in March 1973. Subsequent testing confirmed that cruise missile accuracies would begin at the highest accuracy considered feasible with purely inertially guided weapons.

The most significant ramification of these developments was the pressure put on the Air Force to go all the way in exploiting cruise missile technologies instead of the partial exploitation characterized by the SCAD decoy. As was seen in the latter part of Chapter 4, the fact that the long-range cruise missile clearly manifested U.S. technological superiority became one of the decisive motives for proceeding with development, and certainly strengthened the determination to find a slot for the weapon in the strategic arsenal. There was also a quickening interest, in both the United States and several NATO countries, in tactical and theater applications for cruise missiles, an interest that eventually spawned the GLCM and was in part responsible for the reorientation of the SLCM to a theater role. More generally, as the cruise missile became an accepted weapon within the U.S. military establishment, there was a growing tendency to examine it as a follow-on or modernization option for roles and missions traditionally assigned to aircraft or ballistic missiles.

NOTES

1. In contrast, the nearest Soviet equivalent, the Echo I class SSGNs, which appeared in the same year as the Halibut, sported six SSN-3 launchers and these were promptly followed by Echo IIs with eight launchers.

2. One exception to this statement is of some interest. In 1945 the U.S. Navy was experimenting with a radio-controlled drone powered by a Westinghouse turbojet that was a mere 9.5 inches in diameter (Aviation Week and Space Technology, 5 September 1977, letters to the editor).

3. Arthur J. Alexander and J. R. Nelson, Measuring Technological Change: Aircraft Turbine Engines (Santa Monica, Calif.: Rand Corporation, R-1017-ARPA/PR, May 1972), p. 16.

4. B. Pinkel and J. R. Nelson, A Critique of Turbine Engine Development Policy (Santa Monica, Calif.: Rand Corporation, RM-6100/1-PR, April 1970), p. 12.

5. Ibid., Figure 5, p. 7.

6. Fiscal Year 1972 Authorization for Military Procurement, hearings, p. 3,083.

7. Gordon E. Moore, "Progress in Digital Integrated Electronics," in Basic Limitations in Microcircuit Fabrication Technology, ed. Ivan E. Sutherland, Carver A. Mead, and Thomas E. Everhart (Santa Monica, Calif.: Rand Corporation, R-1956-ARPA, November 1956), Appendix B, p. 35.

8. ATRAN involved matching radar images of the terrain overflown, while TERCOM matches a digitalized picture of the contours of the terrain being overflown. A latter-day version of ATRAN—called Radar Area Correlator—is under development to provide the U.S. Army's Pershing II ballistic missile with a terminal homing capability.

9. John W. R. Taylor, ed., Jane's All the World's Aircraft 1963/1964 (London: Sampson, Low, Marston, 1963), p. 396. The SLAM was under consideration for a MIRV warhead in the late 1950s (Missiles and Rockets, 25 March 1960, p. 12).

10. Flight International (1 October 1977, p. 964) reported that the U.S. electronics company, E-Systems, patented a radar-based, terrain-comparison guidance system in 1958. An E-Systems employee, in a letter to the author, confirmed that TERCOM originated with the SLAM project.

11. Status of U.S. Strategic Power, hearings, p. 27.

12. Fiscal Year 1973 Authorization for Military Procurement, Addendum No. 1, hearings, Senate Armed Services Committee, June-July 1972, p. 4,382.

8
SALT

One of the more established explanations of the decision to develop a strategic cruise missile is that it was intended to serve as a bargaining chip in the SALT II negotiations. As seen in Chapter 3, Secretary of Defense Laird used this argument quite explicitly in justifying his request to initiate development both from the standpoint of bargaining leverage in general and as a precondition for bringing Soviet naval cruise missiles into the purview of SALT. It is quite difficult, however, to reconcile this intention with what transpired in the SALT negotiations up until the signing of the Vladivostok accord in November 1974.

As has been seen (Chapter 3), the question of limiting cruise missiles was raised briefly during the SALT I negotiations. In November 1969, the United States tabled a paper sketching the possible scope of an agreement and one element was that submarine-launched cruise missiles would be limited to those then operational. This U.S. position was subsequently extended to include a ban on the testing and deployment of land-based cruise missiles of intercontinental range. In December 1970—at a time in the negotiations when the two sides were far apart, particularly over the issue of U.S. forward based systems (FBS)—the Soviet Union agreed that the development and deployment of both ground and sea-launched cruise missiles of intercontinental range should be prohibited. According to the chief American negotiator, Gerard Smith, this item simply "got lost in the shuffle" when the ambition for a comprehensive agreement was abandoned.[1]

As far as one can tell from the public record, cruise missiles did not again figure in the SALT discussions until October 1974 during Henry Kissinger's visit to Moscow to prepare for the Vladivostok summit. This is somewhat surprising in view of the visibility of the SLCM program from June 1972 and its quite explicit orientation toward SALT. Although the pre-Vladivostok phase of SALT II can also be described as the pre-cruise missile phase, the subject matter of the negotiations during this period remains relevant to the present discussion. More important, perhaps, was the emergence in the United States of a notably more skeptical attitude toward SALT. Both these factors had a direct impact on the nature and intensity of the controversy over the cruise missile when this subject arose early in 1975.

On the question of U.S. attitudes toward SALT, it is clear that many Americans were troubled by the fact that SALT I permitted the Soviet Union some 40 percent more ICBMs and SLBMs than the United States. In response, a congressional amendment to the declaration authorizing the president to approve the interim agreement on offensive weapons stipulated that future agreements must provide for equality between the United States and Soviet Union in levels of intercontinental strategic forces. Further, a series of developments from 1972 through 1974 served to reinforce the impression in the minds of many people that the United States had been outnegotiated in SALT I, and that the Soviet Union would exploit to the maximum the gaps and ambiguities in the language of the treaty.

An early development concerned whether the Soviet Union had 42 or 48 SSBNs in existence and under construction at the time SALT I was signed. The lower figure was the U.S. intelligence estimate but President Nixon, at the last minute and over the objections of his chief negotiator, agreed to accept the higher number claimed by the Soviet Union. In August 1972 Senator Jackson disclosed that the United States had confirmed the accuracy of its intelligence estimate.[2] Other developments interpreted by some as tests of U.S. resolve and by others as indicative of Soviet disrespect for the spirit of SALT, include the following:

The construction of additional silos in 1973. It was explained that these were intended as launch-control centers but not before the United States had brought the matter up formally before the Standing Consultative Commission;

Employing concealment measures to impede U.S. verification efforts; and

Employing an air defense radar in an ABM mode.[3]

Beyond question, however, the development that most disturbed the United States was the size of the Soviet Union's third-generation ICBMs. Throughout SALT I, a major focus of U.S. concern was the mammoth SS-9 Scarp ICBM. When the treaties were signed it was established that 313 silos for this weapon were completed or under construction, and a clause in the agreement prohibited the replacement of light ICBMs with heavy ICBMs. It was also agreed that, in the process of replacement and modernization, the dimensions of

ICBM silos would not be "significantly increased." Unfortunately, despite persistent U.S. efforts, the Soviet Union refused to agree to a precise definition of light and heavy ICBMs and to what, exactly, would constitute a significant increase in silo dimensions. On the latter issue there was a "common understanding" that the dimensions of existing silos would not be increased by more than 10 to 15 percent, but it was not specified whether this was the permissible increase in depth, diameter, or volume.[4]

In response to this situation, the U.S. delegation opted to issue a unilateral declaration to the effect that "the United States would consider any ICBM having a volume significantly greater than that of the largest light ICBM now operational on either side to be a heavy ICBM."[5] Subsequently, the United States indicated that it understood the clause on the permissible increase in missile size to mean that "the maximum volumetric increase in missile size . . . cannot exceed approximately 32 percent."[6] The Soviet Union violated this unilateral understanding, but the United States chose not to regard it as a breach of the terms of the interim agreement. The Soviet Union's SS-19 ICBM, one of four new ICBMs of which the United States became aware in the latter half of 1973, had an estimated volume of 100 cubic meters compared to 69 cubic meters for the weapon it would replace (the SS-11), a volumetric increase of 45 percent.[7]

Quite apart from its significant military implication (the SS-11 accounted for some 60 percent of the Soviet ICBM force), this development had a profound psychological impact. For most people it became clear why the Soviet Union had been so adamant in its refusal to specifically define light and heavy ICBMs.[8] This development confirmed the view, already well established, that while the Soviet Union would abide by the letter of agreements on arms limitation, it was naive to assume that it was imbued with the spirit of this endeavor.[9]

On the pre-Vladivostok SALT II negotiations themselves, it is now well known that the two sides took extreme positions at the outset and two years of negotiations produced little in the way of reconciliation. It was mutually agreed that the objective for SALT II was a permanent agreement and one that covered weapons not included in SALT I, notably bombers and MIRVed missiles. But, as Paul Nitze described it, the two sides approached SALT II with markedly different interpretations of the meaning of the interim agreement.[10] The interim agreement specifically provided that its provisions were not to prejudice the scope or terms of a permanent agreement. The United States therefore presumed that a comprehensive agreement would be based on the principles of equality or essential equivalence. The Soviet view, on the other hand, was that the unequal missile launcher numbers in the interim agreement, having been granted to compensate for certain geographic and other (primarily technological) inequalities between the two sides, should be carried over essentially unchanged into the permanent agreement.[11]

In addition to this quite fundamental divergence, the Soviet Union insisted (as it had indicated it would in 1971) that because a comprehensive treaty was now under negotiation, the 700-odd U.S. nuclear-capable aircraft deployed in Europe and on aircraft carriers in the European theater should be counted against the U.S. total. The pre-eminent U.S. concern was quite different, namely, the Soviet lead in ICBM launchers and throw-weight and the counterforce potential therein once the Soviet Union acquired a deployable MIRV. This concern understandably increased markedly as the intelligence information came in on the throw-weight of the new ICBMs and when the Soviet Union tested its first MIRV in August 1973.

In the latter months of 1973 the U.S. negotiating position on SALT was in some disarray. There were differences of opinion on whether equality necessitated strict numerical parity in strategic forces.[12] There were also divisions on the question of comparing bombers with ballistic missiles. Some felt the two delivery systems could be viewed as equivalent on a one-for-one basis and that equality in aggregate force levels should be the negotiating target. Others emphasized the differences between strategic delivery vehicles and argued for equality within types, especially ICBMs.[13] Moreover, the two key persons on the U.S. side were heavily preoccupied with other matters: President Nixon with Watergate and Henry Kissinger with Vietnam and the Middle East.

Generally speaking, however, from about the third regular session of SALT II, which began in Geneva on 25 September 1973, the United States endeavored to persuade the Soviet Union to address specifically the question of ICBM throw-weight, particularly that of MIRVed ICBMs. For its part, the Soviet Union professed not to see, or at least refused to concede, that ICBM throw-weight or the number of warheads on ICBMs was in any sense significantly different from other systems. Thus on throw-weight they pointed to the U.S. lead in bomber payload and on warheads they pointed to U.S. superiority in MIRVed SLBMs. In addition, they persisted on the inclusion of FBS on the U.S. side and on general compensation for their geographic and technological disadvantages.

In March 1974 Kissinger went to Moscow to seek new parameters for the negotiations and to overcome the deadlock in the regular sessions in Geneva. He is reported to have made two basic suggestions: first, to negotiate equal aggregate ceilings for the strategic forces with each side free to determine the mix or, second, to develop equality in ICBM throw-weight.[14] Two variations on the latter theme were also offered: the United States would halt its MIRV deployment program if the Soviet Union would limit the number of high throw-weight

MIRVed ICBMs it deployed, and equal ICBM throw-weight could be confined to MIRVed systems with no limits on single-warhead missiles.[15] The Soviets objected to both proposals. They regarded equal aggregate ceilings as unfair in view of the U.S. FBS and the fact that the United Kingdom, France, and China possessed nuclear forces directed against the Soviet Union. The suggestion to develop equal ICBM throw-weight was resisted on the ground that it ignored the U.S. lead on bomber payload. The only counterproposal was an offer by Brezhnev to extend the 1972 interim agreement plus limits on the number of MIRVed missiles that each could deploy. The latter proposal—reported to have been 1,000 MIRVed launchers for the Soviet Union and 1,100 for the United States[16]—reflected the Soviet view that MIRVed missiles should be controlled by number rather than throw-weight. But for the United States, 100 additional launchers was inadequate compensation for the prospective magnitude of the gap in ICBM throw-weight.

At the July 1974 summit in Moscow, the two sides agreed to abandon the ambition of negotiating a permanent agreement in favor of a ten-year (1975 to 1985) agreement. This lower goal did little to reduce the distance between the two sides. President Nixon's proposal included the following:[17]

Mutual limits on the number of MIRVed ICBMs and SLBMs, with the United States granted a higher ceiling to compensate for the greater throw-weight of Soviet missiles. The proposed figures were 1,050 for the United States and between 550 and 700 for the Soviet Union;[18]

Some reductions in the number of single-warhead ICBMs as a move toward greater reliance on SLBMs on the ground that this would reduce mutual concern about a first-strike counterforce threat; and

No restrictions on bombers and SLBMs.

The Soviet counterproposal was a familiar one:[19]

An equal number of MIRVed missiles on each side. Brezhnev reportedly indicated again that the minimum number the Soviet Union would entertain was around 1,000;[20]

The Soviet Union would be permitted to retain its overall lead in numbers of missiles;

The Trident and B-1 programs, being new systems not yet in production, should be stopped; and

FBS should be counted in the U.S. total.

In these circumstances it is scarcely surprising that the fifth regular session of the negotiations (18 September to 5 November 1974) was quite unproductive. The first movement came in October.

Kissinger journeyed to Moscow again with a revised set of SALT proposals and with the knowledge that the impending summit at Vladivostok would achieve nothing if existing differences were not substantially narrowed. Kissinger is reported to have offered the Soviets two broad choices. The first provided for an equal aggregate number of 2,000 central strategic delivery vehicles, including an equal ceiling of 1,000 MIRVed missiles plus sublimits on Soviet heavy ICBMs and U.S. bomber-launched missiles. The second was to offer the Soviets a greater number of non-MIRVed delivery vehicles offset by more MIRVed launchers for the United States.[21] Kissinger found the Soviet posture on SALT a good deal less intractable than in July. As the negotiations progressed, the Soviets expressed interest in the first alternative above.

There has been considerable speculation on the reasons for the flexibility in the Soviet position that emerged between July and October 1974. Some suggest the Soviets had had their own internal controversy on what sort of SALT II treaty was acceptable, and one faction had prevailed or a consensus had been reached. Others were inclined to stress the possibility that the Soviets became convinced that continued stalemate in SALT would threaten detente, and they were willing to bend on the former in order to preserve the latter.[22] Of particular interest here is whether the cruise missile played any part in the process.

The reference to bomber-launched missiles in Kissinger's first proposal was the first U.S. attempt to meet (or perhaps to test) the Soviet argument that the Soviet lead in ICBM throw-weight was matched by the U.S. lead in bomber payload. The question is whether the suggested trade-off between Soviet heavy ICBMs and U.S. bomber-launched missiles was a significant factor in the Soviet interest in the proposal as a whole. It is impossible to offer more than a speculative answer, but it is worth recalling that the first Soviet reaction to the strategic cruise missile programs came in an article in <u>Krasnaya Zvezda</u> on 30 May 1974. That article, devoted primarily to the Harpoon and SLCM missiles, concluded that the U.S. plans for the latter weapon were "by no means just talk."[23] One could tentatively hypothesize that until about mid-1974 the Soviet Union viewed the strategic cruise missile as an unadulterated bargaining chip and refused to be drawn. This is a perfectly reasonable attitude. The Pentagon had not succeeded in persuading Congress that the SLCM made good sense, the Air Force was holding firm to its view that ALCMs should be short-ranged and carried to the target on penetrating bombers, and other launch platforms for cruise missiles (surface ships and ground vehicles) had received only the most cursory examination. But as these programs took hold in the United States, the Soviets began to seriously assess their military implications and concluded they could pose a major new threat. One can also reason in the other direction, namely, the United States during the latter half of 1974 received signals from the Soviet Union that led it to believe that strategic cruise missiles could prove to be more powerful negotiating instruments than had been presumed up to that time.

In any event there was a good deal of optimism that a bargain could be struck at Vladivostok in November 1974. The Soviets had not subscribed to Kissinger's first proposal in its entirety. They indicated a preference for higher ceilings than those proposed by the United States and continued to insist on some compensation for the U.S. FBS and the British and French nuclear forces.[24] But compared to the situation earlier in 1974, the improvement was little short of dramatic.

The broad details of the accord that emerged from Vladivostok are well known. In effect it was a statement on what a ten-year SALT II treaty was intended to look like. As of this date (December 1980) the text of the accord has not been made public, but Thomas Wolfe has pieced together a list of the items agreed upon:[25]

> 1. An equal overall ceiling of 2,400 strategic delivery vehicles for each side, to include ICBMs, SLBMs, and bombers.
> 2. An equal number of 1,320 MIRVed missile launchers for each side, with no limit on throw-weight.
> 3. The counting of any missile tested with MIRV against the MIRV ceiling, if the missile should be deployed.
> 4. Freedom to mix within the agreed aggregate of 2,400 delivery vehicles.
> 5. A sub-limit of 313 on heavy missiles and no new silo construction, provisions to be carried over the Interim Agreement of May 1972.
> 6. Deployment of land-mobile missiles and some types of bomber-launched missiles permitted, but to be included in the overall ceiling of 2,400 delivery vehicles.
> 7. Apparent dropping of the long-standing Soviet demand to account for FBS in any agreed aggregate of central strategic delivery systems.
> 8. No constraints on modernization to preclude such measures as improvements in accuracy and deployment of new systems still under development, e.g., the B-1 bomber and Trident submarine.
> 9. Duration of the new agreement to be from 1975 to 1985, with relevant provisions of the Interim Agreement remaining in force until entry into effect of the new agreement in October 1977.

10. Following conclusion of the new agreement, further negotiations on "possible reductions of strategic arms in the period after 1985" to begin "no later than 1980-1981."

It became apparent very quickly that several key details had been agreed upon only in the broadest terms, and it was acknowledged that substantial negotiations would be required to convert the accord into a formal treaty. However, on the question of bomber-launched missiles (item 6 above), the two sides had conflicting interpretations of what had been agreed. It had been agreed at Vladivostok that the Soviet Union would draft the memorandum on the accord and submit it to the United States for approval. Early in December 1974 there were rumors that the United States was seeking some changes in the Soviet draft, and it can be presumed that these changes included the cruise missile question.[26] Apparently the Soviet view was that the item concerning bomber-launched missiles covered both ballistic and cruise missiles, while the Americans said it only covered the ballistic variety.

The weight of the evidence appears to support the Soviet view. In his December 3 background briefing on the accord, Kissinger listed as one of its "essential elements" the fact that "airborne missiles of a range of more than 600 [kilometers] will be counted as individual missiles, though not as MIRV's."[27] At no point did he specify the type or types of missiles involved, a curious oversight in view of the fact that both countries actually possessed ALCMs with a range in excess of 600 km.—the Soviets the AS-3 Kangaroo and the Americans the AGM-28B Hound Dog. In an editorial on the following day, the New York Times cited what it regarded as essentially a ban on long-range, air-launched ballistic missiles as the single virtue of the accord.[28]

The basis on which this presumption was made is far from clear. There is no question that air-launched cruise missiles were discussed at Vladivostok. Malcolm Currie, the DDR&E at the time, was quite explicit on this: "I can . . . assert that the Soviet Union is aware of these programs and has some deep concerns as expressed to people that were at Vladivostok."[29] Moreover, the ALCM was the only long-range, bomber-launched missile under development in the United States. No long-range, bomber-launched ballistic missile was even being contemplated at the time. The one piece of evidence that supported the U.S. position was that the original offer to limit bomber-launched missiles was made in relation to limits on Soviet heavy ICBMs,[30] and no limitations on these weapons beyond those imposed in SALT I were agreed to at Vladivostok. Nevertheless, since the ALCM was discussed at Vladivostok and since it was the only new long-range bomber weapon in sight, the Soviet interpretation seems entirely justified. What undoubtedly added to the Soviet Union's ire was its perception that it had made very substantial concessions to make the Vladivostok accord possible, especially in agreeing to forego compensation for the British and French nuclear forces and for the U.S. forward-based nuclear weapons.

The point to be stressed here is that largely because of the cruise missile issue, the Soviet Union entered the post-Vladivostok SALT negotiations in a bitter frame of mind. The SALT negotiations are influenced as much by atmosphere and attitudes as by substantive matters. The resentment caused by the split on cruise missiles was a major factor in delaying SALT II, both in itself and because the delay allowed evolving weapons technology to create new obstacles to an agreement.

For the next four-and-one-half years the cruise missile wove a tortuous path through the SALT II negotiations. For a considerable period after Vladivostok, the Soviet Union viewed limitations on cruise missiles more as a question of principle than a subject for give-and-take negotiations. In mid-1975, in addition to holding to its interpretation of the Vladivostok accord on ALCMs, the Soviet Union insisted that all ground- and sea-launched cruise missiles with a range in excess of 600 km. be banned completely. Over time, it became progressively more apparent that, of all the cruise missile variants under development or contemplated in the United States, it was the GLCM that most concerned the Soviets. During 1975 the first intelligence reports appeared on a new Soviet MIRVed IRBM, the SS-20. This development, coming on top of the deployment of a new medium bomber (the Backfire), rekindled concern over the nuclear balance in Europe. The members of NATO, particularly the Federal Republic of Germany, began to urge the United States to take more explicit account of the European strategic balance. Specifically, these countries were increasingly attracted to the GLCM as a near-term option to offset the growth in Soviet theater nuclear capabilities, and sought to prevail upon the United States not to foreclose it in the interests of securing agreement on the central U.S.-Soviet strategic balance.

For reasons that are not difficult to understand, Germany is reserved a special place in Soviet threat perceptions. The prospect of the deployment in this country of long-range, nuclear-armed cruise missiles was undoubtedly the major factor behind the Soviet Union's unwearying insistence that the GLCM and the SLCM be severely constrained in the context of SALT II.[31] The details of the negotiations are not essential to the present study, but the general outcome as far as cruise missiles are concerned was that the United States secured considerable freedom on the deployment of the ALCM, while the protocol to the treaty banned the deployment (but not the testing) of sea- and ground-launched cruise missiles through 31 December 1981.[32]

Bargaining leverage in the forthcoming SALT II negotiations was among the most visible rationales for the decision to initiate development of a strategic cruise missile. This expectation was ultimately fulfilled, probably far more completely than U.S. officials anticipated. Cruise missiles preoccupied the SALT II negotiations at all levels from January 1975 until mid-April 1979, when it was agreed that existing strategic bombers could carry a maximum of 20 ALCMs and that the average number of ALCMs per aircraft (existing strategic bombers and such new carriers as might be developed) could not exceed 28. The curious fact remains, however, that U.S. officials made no attempt to use the cruise missile development program as a negotiating lever for a period of about two years. Indeed, if this account of what transpired at Vladivostok in November 1974 is accurate, U.S. negotiators appear to have overlooked what was being done as far as cruise missile development was concerned. Essentially, it would seem that it was the Soviet reaction to these developments that belatedly persuaded U.S. officials that the cruise missile could be a lever of some consequence in the negotiations.

NOTES

1. Gerard Smith, Doubletalk: The Story of SALT I (New York: Doubleday, 1980), p. 190. See also pp. 89 and 130.
2. The Age (Melbourne), 17 August 1972, p. 7.
3. For more details on these and other issues concerning compliance with the terms of SALT I, see The State Department, Selected Documents, No. 7, 1978.
4. The text of the Interim Agreement together with agreed interpretations and common understandings is reprinted in John H. Barton and Lawrence D. Weiler, eds., International Arms Control: Issues and Agreements (Stanford, Calif.: Stanford University Press, 1976), pp. 375-82.
5. Ibid., p. 381.
6. Military Implications of the Strategic Arms Limitation Talks Agreements, hearings, House Armed Services Committee, July 1972, pp. 15, 144-45.
7. See Colin S. Gray, "SALT II and the Strategic Balance," British Journal of International Studies 1 (1975): 190.
8. It appears that the U.S. delegation was informally advised at Helsinki in May 1972 that the Soviet Union would be deploying new ICBMs in SS-11 silos, but that these new missiles would not exceed the halfway mark in terms of volume between the SS-11 and the SS-9. However, to the best of my knowledge, this fact was not made generally known until January 1976. See the article by Gerard E. Smith, the

chief U.S. negotiator during SALT I, in the New York Times, 16 January 1976, p. 29.

9. The Carter administration had evidently taken this lesson to heart, at least as far as unilateral declarations are concerned. In July 1978 Paul C. Warnke assured a conference audience that there would be no such declarations in SALT II.

10. Nitze was on the U.S. negotiating team until June 1974.

11. Paul H. Nitze, address before the staff of the Los Alamos Scientific Laboratory, 17 December 1974, reprinted in Hearings on Military Posture and H.R. 3689, House Armed Services Committee, February 1975, pp. 1,617-43.

12. Times (London), 29 December 1973, p. 11.

13. New York Times, 6 February 1974, pp. 1, 12.

14. New York Times, 24 March 1974, p. 3.

15. New York Times, 12 April 1974, p. 2.

16. New York Times, 26 December 1974 (editorial).

17. New York Times, 9 July 1974, p. 1.

18. Ibid., p. 3.

19. Ibid., p. 1.

20. New Yorker, 29 July 1974, p. 70.

21. New York Times, 3 December 1974, p. 1.

22. On these respective speculations see Thomas W. Wolfe, The SALT Experience: Its Impact on U.S. and Soviet Strategic Policy and Decisionmaking (Santa Monica, Calif.: Rand Corporation, R-1686-PR, September 1975), pp. 161-64.

23. Quoted in Defense/Space Business Daily, Washington, D.C., 17 June 1974, p. 256.

24. "Text of Secretary of State Henry A. Kissinger's Background Briefing on Vladivostok, 3 December 1974," in Nuclear Strategy and National Security: Points of View, ed. Robert J. Pranger and Roger P. Labrie (Washington, D.C.: American Enterprise Institute for Public Policy Research, 1977), p. 399.

25. Wolfe, The SALT Experience, pp. 165-66.

26. New York Times, 10 December 1974, p. 12.

27. Pranger and Labrie, Nuclear Strategy and National Security: Points of View, p. 397.

28. New York Times, 4 December 1974 (editorial). For the rest, the editorial was highly critical because it permitted virtually all the known plans on each side for the expansion and modernization of the strategic force. One of the best elaborations of this line of criticism is Milton Leitenberg, "The Vladivostok Ceilings, and Why They are So High," British Journal of International Studies (July 1976): 149-63.

29. Fiscal Year 1976 and July-September 1976 Transitional Period Authorization for Military Procurement, hearings, Senate Armed Services Committee, February-April 1975, p. 5,180.

30. In his background briefing Kissinger indicated that these two issues were again discussed jointly at Vladivostok. See Pranger and Labrie, Nuclear Strategy and National Security: Points of View, p. 400. It can also be mentioned that on 24 October 1974, the U.S. Air Force conducted a test involving the air launch of a Minuteman I ICBM. However, this test was connected with the interest in mobile ICBMs; it was not an air-launched ballistic missile in the sense relevant here.

31. The intensity of the Soviet feelings about these weapons was revealed in its sharp and uncompromising reaction to the NATO decision in December 1979 to begin deployment in 1983 of 464 GLCMs and 108 extended-range Pershing IIs.

32. For a detailed account of the role played by cruise missiles in SALT II, see the author's The Cruise Missile and Arms Control, Canberra Papers on Strategy and Defence, No. 20, Australian National University Press, 1980.

9
CONCLUSIONS

Viewed from a general perspective, the history of the long-range cruise missile is by no means atypical. As with other major weapon systems, the total explanation of its development history is multifaceted and no one strand of explanation remained dominant throughout. Nevertheless, the results of earlier investigations into the origins of major weapon systems cannot be generalized: the cruise missile was the product of its own time and has its own story.

As regards the main purpose of this inquiry—the reasons for the sudden decision to develop a strategic submarine-launched cruise missile*—the foregoing analysis suggests that the standard explanations are less than satisfactory. Technological determinism—the belief that it is technically feasible to develop a new weapons system plus strong institutional pressures to proceed just because it is feasible—played a relatively minor role. In fact, it appears that development of the SLCM could have been initiated around the mid-1960s with about the same degree of success as has been experienced with the current program. Subsequently, when it became apparent that the cruise missile would be a supremely elegant weapon and one that clearly reflected U.S. superiority in the relevant technologies, there emerged a strengthening determination to find a slot for it in the strategic arsenal.

---

*It is true that there were some efforts to push the Air Force in the direction of a long-range, nuclear-armed cruise missile as early as 1967-68. At the time, however, the proponents of such a weapon were few in number and, in general, their proposals were nonspecific and long term in character. In contrast, the proposal to develop the SLCM came directly from the secretary of defense along with a request for $20 million, a relatively large sum for the initiation of a new weapons system.

Institutional pressure to develop this type of weapon was conspicuous largely by its absence. The Navy was at best indifferent toward strategic cruise missiles, while the Air Force was positively against them. Within the defense bureaucracy, support for the SLCM seems, initially, to have been confined to a small group of technically oriented civilians in the Office of the Secretary of Defense, particularly the research and development division.

The action-reaction theorem is of limited value in explaining the emergence of the SLCM. It is true that there was no lack of concern about the strategic balance and the pace and scope of the Soviet strategic buildup. But no attempt was made to present the SLCM as the most effective response to a particular new Soviet strategic weapon or strategic capability. The predominant U.S. concern was the number and size of Soviet ICBMs and the future counterforce potential of these weapons. In 1972 (and for several years thereafter) no one in the United States viewed the long-range cruise missile as an appropriate offsetting augmentation of U.S. strategic capabilities. The only specific threat mentioned in connection with the original SLCM proposal was the strategic potential of Soviet naval tactical cruise missiles. However, the evidence would support the conclusion that this threat was more of an excuse than a rationale for the SLCM, a supplementary argument on which to fall back in the event that the proposal was criticized as having no concrete foundation.

Another possibility to be considered is that the long-range cruise missile filled a void in U.S. strategic capabilities that revisions in strategic doctrine had opened up. Although the decisive changes in declaratory policy did not occur until 1974 under James Schlesinger, the considerations the United States declared to be the basis on which it determined the strategic capabilities required had been undergoing important revisions since 1969. These revisions fell into two broad categories: refinement of the concept of assured destruction in the form of an increasing emphasis on insuring that the Soviet Union would not recover from a nuclear war more rapidly than the United States; and a movement toward increasing the number of ways in which strategic weapons could be used.

It was suggested above that the cruise missile is well suited to meet the assured destruction/war recovery requirement. The relevant targets are not time-urgent, so the cruise missile's low speed is not a handicap. Its high accuracy and relatively large warhead make it more than a match for even the hardest industrial and economic targets. As regards the quest for more strategic options, it has been seen that the United States was moving far more rapidly in this direction at the level of weapon development and strategic targeting plans than its declared policy indicated. In short, the United States was moving decisively away from the assured destruction second strike as its principal strategic option at the time the strategic cruise missile was proposed.

Once again, however, it seems one must look elsewhere for an explanation of the proposal to develop a strategic cruise missile. It was never suggested that the strategic forces in existence in 1972 lacked a particular capability deemed necessary to deterrence and that the cruise missile was intended to fill the gap. John Foster's forceful defense of the proposal was a plea for yet more diversity in the strategic forces, not a claim that the cruise missile could do something existing forces could not do. It was not until 1975 that a doctrine-related rationale was provided for the SLCM, namely, "a unique potential for unambiguous, controlled single-weapon response." This rationale seemed unconvincing and was soon dropped.

The remaining possibility is that the cruise missile was conceived essentially as a bargaining chip in the context of the SALT negotiations. In itself this is not particularly persuasive because the United States displayed no inclination to exploit the development of this weapon as a tactical lever in the negotiations until the Soviet Union revealed its concern. Even then it was very nearly given away.

It can be argued, however, that the cruise missile was a bargaining chip of a far more profound sort. During the negotiations on SALT I it became apparent that the mutual interest in reducing the risk of strategic war in no way compensated for the deep historical antagonism between the two parties and the competititve instincts this antagonism had bred. The negotiation table became another arena in which the competition was carried on. There can be little doubt that the military community and its sympathizers in the United States were profoundly disturbed by the numerical imbalances accepted by the political leadership in SALT I. The Office of the Secretary of Defense prepared a list of suggestions for Secretary Laird on the strategic programs that could be accelerated and "looking at the situation in the light of the [SALT] agreements," Laird picked the strategic cruise missile.[1] From this perspective the SLCM proposal might be characterized as a reflex action on the part of the military community to protest the severity of the constraints imposed on the United States relative to those imposed on the Soviet Union.*

---

*It was seen in Chapter 3 that one of the explicit reasons advanced for proceeding with the SLCM was the ambition to compensate somewhat for the numerical inferiority in ballistic missiles imposed on the United States in SALT I. Such a view is supported by Kissinger, who sometimes characterized the SALT I agreement as a five-year

To pursue this line of reasoning, the proposal to fit a cruise missile into the strategic arsenal appears to have been an outgrowth of the perception that the United States had lost the initiative both in the strategic nuclear competition and in foreign policy generally. By the early 1970s the United States had been emotionally and physically drained by the war in Indochina. In the negotiations on SALT I, U.S. officials perceived themselves to be on the defensive, reduced to watching the Soviet Union increase its lead in numbers of ICBMs and rapidly close the gap on SLBMs. The fact that the United States was widening its lead in the number of warheads appears, psychologically at least, to have been inadequate compensation. The Soviet strategic program was steaming along with both production and development in high gear while the United States would have nothing beyond MIRV to put into production for at least five years.

Thus development of the B-1 strategic bomber was started in 1970 followed by the Trident SLBM in 1971, with the latter being put on an accelerated schedule in the following year. However, it appears that a small group of key officials judged that this was not enough, that more evidence was needed to demonstrate that the United States was not down and out as a result of Vietnam and not intimidated by the momentum of the Soviet strategic program, but capable of and prepared to compete with vigor and imagination.

Just why the cruise missile was selected to provide this additional evidence is a difficult question to answer. It may have been through a process of elimination tempered by the judgment that Congress was in no mood to entertain ambitious military schemes and higher budgets for the Pentagon. Follow-on systems for the bomber and SLBM legs of the Triad were under development, and production of the Minuteman III ICBM and the Poseidon SLBM was in full swing. Initiating the development of a new ICBM was probably judged to be unacceptable to Congress and, in any case, the terms of SALT I (as unilaterally interpreted by the United States) restricted such a weapon to an increase in volume of 30 percent over the Minuteman III and to deployment in fixed silos. The cruise missile, on the other hand, was not affected by SALT, did not duplicate an existing program, and could be presented as a distinctly modest initiative, a couple of hundred weapons deployed on ten old SSBNs or thinly distributed over a larger number of SSNs.

In sum, a plausible but admittedly speculative answer to this

---

opportunity for the United States to catch up strategically. See his White House Years (Boston: Little, Brown, 1979), p. 1,245.

basic question is that the proposal to develop the strategic cruise missile was, in essence, psychologically motivated: to strengthen the signal to the Soviet Union that the United States would vigorously contest any Soviet bid for strategic superiority, and to help alleviate internal anxieties that the United States had lost its self-confidence and the will to compete.

The United States is now committed to the deployment initially of some 3,400 strategic cruise missiles, perhaps the largest procurement of a single type of strategic weapon in its history. These weapons, however, will be launched from B-52 strategic bombers, not from submarines as originally envisaged. From a strictly military viewpoint the ALCM is undoubtedly an extremely cost-effective program, but the manner in which the United States was led to this action is cause for considerable disquiet as to the rationality of the weapons acquisition process. First of all, the switch in launch platforms lends additional support to the contention that the original SLCM proposal was a hasty gut reaction to the circumstances in which the United States found itself in the early 1970s. In addition, however, the United States was led to the ALCM by a process that was dominated by negative reasoning. For a critical period of time there was a high degree of coincidence between the individuals and groups that supported a long-range (standoff) ALCM and those who opposed the B-1. Moreover, support for the ALCM derived from opposition to the B-1, not the other way around. That cruise missiles will form a major component of the U.S. strategic arsenal in the 1980s could be said to have been due, to an important extent, less to support for this weapon in its own right than to its central role in the campaign to stop production and deployment of a new penetrating strategic bomber.

As is nearly always the case with a conceptually new weapons system, the cruise missile has had and undoubtedly will continue to have significant ramifications. This is not the place to dwell at length on this issue, but two implications would appear to be both conspicuous and important: the impact on SALT on the one hand and on the dynamics of the strategic nuclear arms race on the other.

The U.S. view that verification difficulties would probably permanently exclude cruise missiles from accountability in SALT was not shared by the Soviet Union.[2] The latter country's unswerving determination to have U.S. cruise missiles constrained by SALT led to the undermining of the main organizing principle on which the negotiations had been based, namely, the ability to identify a class of weapons in terms of its (strategic) function. Cruise missiles are a problem because they look the same whether they can travel 300 miles or 2,000 miles, are accurate enough to be militarily attractive with either a nuclear or conventional warhead, and are so compact and adaptable that they can be deployed on just about any platform. In other words, the characteristics of a cruise missile are such that its range and type of warhead—and therefore, its role—cannot be reliably deduced by external inspection, monitoring test flights, or noting the platform on which it is deployed. Since these three procedures constitute the core of prevailing verification capability, the widespread deployment of the cruise missile in its several variants can be expected to seriously jeopardize this cornerstone of strategic arms negotiations. Furthermore, the advent of the cruise missile has irreversibly expanded the scope of SALT to include theater strategic weapons. SALT III—should we be fortunate enough to see it begin—will face the task of determining anew what is to be accountable and what can be left out. Given the diversity of systems with theater strategic capabilities, the number of interested parties, and geographic asymmetries, this task will not be an easy one.

The advent of the modern cruise missile will greatly intensify the strategic competition in bombers, bomber weapons, and bomber defenses. This has been a relatively quiet corner of the strategic competition for nearly 20 years, but the Soviet Union is now close to the deployment of at least one new strategic bomber and is developing a new ALCM of its own. The first tests of the latter weapon were observed late in 1978, which suggests that development was initiated in 1975 or 1976. Whatever the capabilities of these weapons, their impact on the U.S. strategic program is likely to be substantial as, at the present time, the continental United States is well-nigh defenseless against bomber attack. Furthermore, the impact of the cruise missile— if its accuracy and invulnerability live up to expectations—is likely to extend beyond the bomber realm. As regards the crisis stability of the nuclear balance, it has been customary to give the cruise missile high marks because of its slow speed and consequent inability to pose a threat of a surprise first strike. On the other hand, it could be argued that the central consideration in determining whether a weapon does or does not have a first-strike capability is the defender's confidence in being able to ride out the attack and emerge with enough surviving weapons to retaliate. If the attack is made by some 2,500 ALCMs each carrying a 200-KT warhead to within 300 feet or better of the target, the defender's inclination to withhold all or at least some of its weapons will clearly be very low. It is true that if a nuclear exchange did take place bombers and bomber-launched weapons would not be the first to arrive on the targets; but this does not alter the fact that the more certain it is that an attacking weapon will destroy its target in a counterforce strike the stronger will be the pressure to use all the weapons under attack. In sum, the impact of cruise missile deployment on the crisis stability of the strategic balance may well be negative rather than positive unless highly effective defenses against it become available. This does not seem likely in the foreseeable future.

The long-range, nuclear-armed cruise missile is not yet deployed but, in the author's judgment, it is already clear that this is another genie that should have been kept in its bottle.

NOTES

1. The quotation is taken from congressional testimony by John Foster, Jr.; see <u>Fiscal Year 1973 Authorization for Military Procurement, Addendum No. 1: Amended Military Authorization Request Related to Strategic Arms Limitation Agreement</u>, hearings, Senate Armed Services Committee, June-July 1972, p. 353. The term "accelerate" is technically correct in view of the strategic element added previously to the Navy's advanced cruise missile program. In effect, however, it was a new program.

2. Early in 1974 the Navy included in an official list of reasons for proceeding with the SLCM: the fact that the weapon could be covertly deployed; and the fact that the identity of strategic and tactical versions could not be verified (<u>Fiscal Year 1975 Authorization for Military Procurement</u>, hearings, Senate Armed Services Committee, April 1974, p. 3,620). A year later, Chief of Naval Operations Admiral Holloway testified that any limitations on sea-launched cruise missiles, other than a total ban on both tactical and strategic variants, was likely to be meaningless (<u>Hearings on Military Posture and H.R. 3689</u>, House Armed Services Committee, February-May 1975, pp. 840-41).

EPILOGUE

The main purpose of this book has been to account for the U.S. decision to reintroduce the cruise missile to its strategic nuclear arsenal. The commitment to do this became essentially irreversible in June 1977 with the cancellation of the B-1 strategic bomber. Subsequent developments in the cruise missile field have been touched upon somewhat eclectically, so a brief description of the situation as it now stands (December 1980) is in order.

The program for the AGM-86B strategic ALCM remains on schedule. The first B-52G with 12 external ALCMs will be operational in September 1981, and the first complete squadron of 14 aircraft will be ready in December 1982, the official date for initial operational capability. For the next seven years 480 ALCMs will be produced annually with delivery of the last weapon scheduled for May 1989. Installation of a new rotary launcher to permit B-52Gs to carry up to eight additional ALCMs internally is scheduled to begin in October 1985.

The GLCM is also reasonably secure. The first operational test flight was conducted early in 1980 and IOC is planned for December 1983. A total production run of 560 missiles is envisaged to support the planned deployment of 464 weapons. Given the reservations expressed by several European NATO countries—particularly Belgium and the Netherlands—it is possible that the distribution of GLCMs may be revised. Similarly, on 17 October 1980, the United States and the Soviet Union opened negotiations on long-range theater nuclear weapon systems. It is at least conceivable that these negotiations will lead to some delay and perhaps even diminution of GLCM deployment.

The basic SLCM—now called the TLAM-N (Tomahawk land-attack missile-nuclear)—remains on a leisurely development schedule. No date has been set for a production decision and, therefore, there is no target IOC. The deployment figure cited for planning purposes is 125 missiles. The picture is much more defined, however, with regard to SLCMs armed with conventional warheads (but in outward appearance identical to the nuclear version). The TASM (Tomahawk antiship missile) is scheduled for IOC on Los Angeles-class SSNs in June 1982 and on Spruance-class destroyers in June 1983. Total initial procurement of this variant is 243 missiles. A land-attack variant with a conventional warhead (TLAM-C) is scheduled for IOC on submarines in January 1982 and on destroyers in June 1983. This version will have, in addition to TERCOM, a terminal homing system to provide the requisite accuracy. A total of 71 missiles is planned.

Both the TASM and the TLAM-C have been incorporated into the vertical launch system (VLS) program. The VLS program is intended to provide major surface warships (DD963, CG-47, and the proposed DDGX) with a single compact launcher that can fire antiaircraft, antiship, and antisubmarine missiles (the SM-2 Standard, Harpoon, TASM, TLAM-C, and ASROC).

Several other applications for the current cruise missile are under active consideration. Perhaps the most imminent of these is a tactical, medium-range, air-to-surface missile (MRASM) for joint Navy-Air Force use. The MRASM would have a maximum range of about 480 km. and would be designed for launch from such aircraft as the A-6E, A/L-18, P-3, F-16, and the F-111.*

Finally, second-generation strategic ALCMs are already under investigation. The development effort will focus on engines and fuels to increase range up to 2,600 n.m. without increasing the size of the missile. Extensive use will also be made of stealth technologies to reduce radar cross section and infrared signatures. The development schedule envisages test flights in 1983 and an IOC in 1986-87.

---

*If the range of this weapon exceeded 600 km., all carrier aircraft would become strategic bombers under the terms of SALT II.

APPENDIX

U.S. Congress, Senate, Preparedness Investigating Subcommittee of the Committee on Armed Services, Enquiry into Satellite and Missile Programs, 85th Congress, 2d sess., 1958.

U.S. Congress, House of Representatives, Committee on Armed Services, Investigation of National Defense Missiles, 85th Congress, 2d sess., 1958.

U.S. Congress, Senate, Preparedness Investigating Subcommittee of the Committee on Armed Services, Major Defense Matters, 86th Congress, 1st sess., 1959.

U.S. Congress, Senate, joint hearings before the Preparedness Investigating Subcommittee of the Committee on Armed Services and the Committee on Aeronautical and Space Sciences, Missile and Space Activities, 86th Congress, 1st sess., 1959.

U.S. Congress, House of Representatives, Committee on Armed Services, Hearings on Military Posture and H.R. 4016, 89th Congress, 1st sess., 1966.

U.S. Congress, Senate, Committee on Armed Services and Subcommittee on the Department of Defense of the Committee on Appropriations, Military Procurement Authorization for Fiscal Year 1968, 90th Congress, 1st sess., 1967.

U.S. Congress, House of Representatives, Committee on Armed Services, Hearings on Military Posture and a Bill (H.R. 9240), 90th Congress, 1st sess., 1967.

U.S. Congress, Senate, Committee on Armed Services, Authorization for Military Procurement, Research and Development, Fiscal Year 1969, and Reserve Strength, 90th Congress, 2d sess., 1968.

U.S. Congress, House of Representatives, Committee on Armed Services, Hearings on Military Posture and an Act (S. 3293), 90th Congress, 2d sess., 1968.

U.S. Congress, Senate, Preparedness Investigating Subcommittee, Committee on Armed Services, Status of U.S. Strategic Power, 90th Congress, 2d sess., 1968.

U.S. Congress, Senate, Committee on Armed Services, Authorization for Military Procurement, Research and Development, Fiscal Year 1970, and Reserve Strength, 91st Congress, 1st sess., 1969.

U.S. Congress, Senate, Committee on Armed Services, Authorization for Military Procurement, Research and Development, Fiscal Year 1971, and Reserve Strength, 91st Congress, 2d sess., 1970.

U.S. Congress, House of Representatives, Committee on Armed Services, Hearings on Military Posture and Legislation to Authorize Appropriations During the Fiscal Year 1970 for Procurement of Aircraft, Missiles, Naval Vessels, and Tracked Combat Vehicles, Research, Development, Test and Evaluation for the Armed Forces, and to Prescribe the Authorized Strength of the Reserve Forces and for Other Purposes, 91st Congress, 1st sess., 1969.

U.S. Congress, House of Representatives, Committee on Armed Services, Hearings on Military Posture and Legislation to Authorize Appropriations during the Fiscal Year 1971 for Procurement of Aircraft, Missiles, Naval Vessels, and Tracked Combat Vehicles, and Other Weapons, and Research, Development, Test, and Evaluation for the Armed Forces, and to Prescribe the Authorized Personnel Strength of the Selected Reserve of, each Reserve Component of the Armed Forces, and for Other Purposes, 91st Congress, 2d sess., 1970.

U.S. Congress, Senate/House, Committees on Armed Services, CVAN-70 Aircraft Carrier, 91st Congress, 2d sess., 1970.

U.S. Congress, Senate, Committee on Armed Services, The Limitation of Strategic Arms, 91st Congress, 2d sess., 1970.

U.S. Congress, Senate, Committee on Armed Services, Fiscal Year 1972 Authorization for Military Procurement, Research and Development, Construction and Real Estate Acquisition for the Safeguard ABM, and Reserve Strength, 92d Congress, 1st sess., 1971.

U.S. Congress, House of Representatives, Committee on Armed Services, Hearings on Military Posture and H.R. 3818 and H.R. 8687, 92d Congress, 1st sess., 1971.

U.S. Congress, Senate, Committee on Armed Services, Fiscal Year 1973 Authorization for Military Procurement, Research and Development, Construction Authorization for the Safeguard ABM, and

Active Duty and Selected Reserve Strengths, 92d Congress, 2d sess., 1972.

U.S. Congress, Senate, Committee on Armed Services, Fiscal Year 1973 Authorization for Military Procurement, Research and Development, Construction Authorization for the Safeguard ABM, and Active Duty and Selected Reserve Strengths; Addendum No. 1: Amended Military Authorization Request Related to Strategic Arms Limitation Agreement, 92d Congress, 2d sess., 1972.

U.S. Congress, House of Representatives, Committee on Armed Services, Hearings on Military Posture and H.R. 12604, 92d Congress, 2d sess., 1972.

U.S. Congress, Senate, Committee on Foreign Relations, Strategic Arms Limitation Agreements, 92d Congress, 2d sess., 1972.

U.S. Congress, Senate, Committee on Armed Services, Military Implications of the Treaty on the Limitation of Anti-ballistic Missile Systems and the Interim Agreement on Limitation of Strategic Offensive Arms, 92d Congress, 2d sess., 1972.

U.S. Congress, House of Representatives, Committee on Armed Services, Military Implications of the Strategic Arms Limitation Talks Agreements, 92d Congress, 2d sess., 1972.

U.S. Congress, Senate, Committee on Armed Services, Fiscal Year 1974 Authorization for Military Procurement, Research and Development, Construction Authorization for the Safeguard ABM, and Active Duty and Selected Reserve Strengths, 93d Congress, 1st sess., 1973.

U.S. Congress, House of Representatives, Committee on Armed Services, Hearings on Cost Escalation in Defense Procurement Contracts and Military Posture and H.R. 6722, 93d Congress, 1st sess., 1973.

U.S. Congress, Senate, Armed Services Committee, Nominations of Mendolia, McClay, Currie and Bowers, 93d Congress, 1st sess., 1973.

U.S. Congress, Senate, Committee on Armed Services, Nominations of McLucas and Brown, 93d Congress, 1st sess., 1973.

U.S. Congress, Senate, Committee on Armed Services, *Fiscal Year 1975 Authorization for Military Procurement, Research and Development, and Active Duty, Selected Reserve, and Civilian Personnel Strengths*, 93d Congress, 2d sess., 1974.

U.S. Congress, House of Representatives, Committee on Armed Services, *Hearings on Military Posture and H.R. 12564*, 93d Congress, 2d sess., 1974.

U.S. Congress, House of Representatives, Committee on Appropriations, *Department of Defense Appropriations for 1975*, 93d Congress, 2d sess., 1974.

U.S. Congress, Senate, Committee on Foreign Relations, *Briefing on Counterforce Attacks*, 93d Congress, 2d sess., 1974.

U.S. Congress, Senate, Committee on Foreign Relations, *U.S.-U.S.S.R. Strategic Policies*, 93d Congress, 2d sess., 1974.

U.S. Congress, Senate, Committee on Foreign Relations, *Nuclear Weapons and Foreign Policy*, 93d Congress, 2d sess., 1974.

U.S. Congress, Senate, Committee on Foreign Relations, *Detente*, 93d Congress, 2d sess., 1974.

U.S. Congress, Senate, Committee on Armed Services, *Fiscal Year 1976 and July-September Transitional Period Authorization for Military Procurement, Research and Development, and Active Duty, Selected Reserve, and Civilian Personnel Strengths*, 94th Congress, 1st sess., 1975.

U.S. Congress, House of Representatives, Committee on Armed Services, *Hearings on Military Posture and H.R. 3089*, 94th Congress, 1st sess., 1975.

U.S. Congress, House of Representatives, Committee on Appropriations, *Department of Defense Appropriations for 1976*, 94th Congress, 1st sess., 1975.

U.S. Congress, House of Representatives, Committee on International Relations, *The Vladivostok Accord: Implications to U.S. Security, Arms Control, and World Peace*, 94th Congress, 1st sess., 1975.

U.S. Congress, Senate, Committee on Foreign Relations, *Effects of Limited Nuclear Warfare*, 94th Congress, 1st sess., 1975.

APPENDIX / 199

U.S. Congress, House of Representatives, Committee on Armed Services, Overall National Security Programs and Related Budget Requirements, 94th Congress, 1st sess., 1975.

U.S. Congress, Senate, Committee on Armed Services, Fiscal Year 1977 Authorization for Military Procurement, Research and Development, and Active Duty, Selected Reserve, and Civilian Personnel Strengths, 94th Congress, 2d sess., 1976.

U.S. Congress, House of Representatives, Committee on Armed Services, Hearings on Military Posture and H.R. 11500, 94th Congress, 2d sess., 1976.

U.S. Congress, House of Representatives, Committee on Appropriations, Department of Defense Appropriations for 1977, 94th Congress, 2d sess., 1976.

U.S. Congress, House of Representatives, Committee on the Budget, Fiscal Year 1977 Defense Budget, 94th Congress, 2d sess., 1976.

U.S. Congress, House of Representatives, Committee on International Relations, First Use of Nuclear Weapons: Preserving Responsible Control, 94th Congress, 2d sess., 1976.

U.S. Congress, House of Representatives, Committee on International Relations, US-USSR Relations and Strategic Balance, 94th Congress, 2d sess., 1976.

U.S. Congress, Senate/House of Representatives, Joint Committee on Defense Production, Conflict of Interest and the Condor Missile Program, 94th Congress, 2d sess., 1976.

U.S. Congress, Senate, Committee on Armed Services, Fiscal Year 1978 Authorization for Military Procurement, Research and Development, and Active Duty, Selected Reserve, and Civilian Personnel Strengths, 95th Congress, 1st sess., 1977.

U.S. Congress, House of Representatives, Committee on Armed Services, Hearings on Military Posture and H.R. 5068 [H.R. 5970], 95th Congress, 1st sess., 1977.

U.S. Congress, House of Representatives, Committee on Appropriations, Department of Defense Appropriations for 1978, 95th Congress, 1st sess., 1977.

U.S. Congress, House of Representatives, Committee on Armed Services, Hearings on H.R. 8390 and Review of the State of U.S. Strategic Forces, 95th Congress, 1st sess., 1977.

U.S. Congress, Senate, Committee on Foreign Relations, United States/Soviet Strategic Options, 95th Congress, 1st sess., 1977.

U.S. Congress, Senate, Committee on Foreign Relations, Briefings on SALT Negotiations, 95th Congress, 1st sess., 1977.

U.S. Congress, Senate, Committee on Foreign Relations, Warnke Nomination, 95th Congress, 1st sess., 1977.

U.S. Congress, Senate, Committee on Armed Services, Consideration of Mr Paul C. Warnke to be Director of the U.S. Arms Control and Disarmament Agency and Ambassador, 95th Congress, 1st sess., 1977.

U.S. Congress, House of Representatives, Committee on International Relations, The Soviet Union: Internal Dynamics of Foreign Policy, Present and Future, 95th Congress, 1st sess., 1977.

U.S. Congress, Senate, Committee on Armed Services, Department of Defense Authorization for Appropriations for Fiscal Year 1979, 95th Congress, 2d sess., 1978.

U.S. Congress, Senate, Committee on Armed Services, Fiscal Year 1978 Supplemental Military Authorization, 95th Congress, 2d sess., 1978.

U.S. Congress, House of Representatives, Committee on Armed Services, Hearings on Military Posture and H.R. 10929, 96th Congress, 1st sess., 1978.

INDEX

ABM, 34-35
aircraft: AWACS, 70, 84, 125; B-1, 5-6, 11, 12, 34, 36, 40, 45, 66, 70-71, 72, 74-75, 77, 78, 81-83, 189, 190; B-52, 4, 11, 12, 19, 44, 63-64, 70-71, 75, 77-79, 81-82, 107, 190; B-70, 62-63; Backfire, 48, 107, 182; cruise missile carrier, 73, 82, 123-24, 126
aircraft carriers, 29

Brezhnev, Leonid, 42, 178
Brown, Harold, 51, 80, 85-86, 116, 117-18, 121, 128, 138, 146, 154-55, 156-57, 159
Brzezinski, Zbigniew, 154

Carter, Jimmy, 78, 80, 154
civil defense, 145-47
Currie, Malcolm, 40-41, 42, 43, 45, 46, 72, 158, 181

first strike, 118-24, 126, 127, 139, 156
Ford, Gerald, 42
forward based systems (FBS), 177, 178, 180, 182
Foster, John, Jr., 34-36, 37, 65-66, 69, 157-58, 171, 188

Garwin, Richard, 36
Gouré, Leon, 145, 146

Joint Strategic Bomber Study, 74, 76, 80-81

Kennedy, J. F., 136
Kissinger, Henry, 34, 121, 141, 174, 177, 179, 181

Laird, Melvin, 31, 32, 33-34, 154, 188
Locke, Walter, 39, 41, 43, 47, 48

McNamara, Robert, 62-63, 64-65, 136-37
missiles, cruise: advanced cruise missile, 30-31; ALCM, 3-4, 5, 6, 7, 8, 9, 12, 44-45, 49, 51, 62, 71-86, 123-27, 128, 156, 159, 182, 190, 191, 193; ASALM, 75, 86, 125; AS-3 Kangaroo, 181; Crossbow, 61; GLCM, 3, 4, 49-53, 172, 182, 193; Goose, 62; Harpoon, 29-30, 42, 101, 179; Hound Dog, 18-20, 61, 167, 168, 181; JB-2 Loon, 15; Longbow, 61; Mace, 23-24, 167, 168; Matador, 21-22, 23, 167, 168; Navaho, 17-18; Quail, 62, 65, 168; Regulus I, 20, 21, 167, 168; Regulus II, 20-21, 166, 167, 168; SLCM, 3, 4, 6, 7, 8, 9, 31-53, 71, 75, 128, 172, 179, 187, 188, 189, 193; Snark, 16-17; SSN-3 Shaddock, 34, 101, 102; V-1, 15

missiles, other: Exocet, 3; Minuteman III, 12, 105, 111, 149-50, 155, 189; MX, 11, 12, 127, 153, 155; Pershing I, 24; Pershing II, 51-53; Polaris, 20-21, 31, 106-7; Poseidon, 106-7, 109, 139, 189; Rascal, 61; SA-10, 84-85; Skybolt, 62; SRAM, 64, 76; SS-6 Sapwood, 104; SS-9 Scarp, 104-5, 175; SS-11 Sego, 104-5; SS-16, 105; SS-18, 105, 109, 111; SS-19, 105, 111, 117, 176; SS-20, 105, 182; SSN-2 Styx, 3, 29; Trident, 11, 12, 31, 107, 189; Wagtail, 61

National Security Defense Memorandum 242 (NSDM-242), 154
Nitze, Paul, 119-21
Nixon, Richard, 138, 141, 175, 177, 178

Perry, William, 51, 80, 83, 85
Pipes, Richard, 146
Presidential Review Memorandum 10 (PRM-10), 146, 154
Proxmire, William, 36, 69

remotely piloted vehicles (RPV), 168-69
Richardson, Eliot, 36-37
Rumsfeld, Donald, 45, 153, 155

SALT I, 31, 33-34, 36, 117, 138, 142, 174, 175, 189
SALT II, 45, 83, 119, 126, 155, 157, 174-83

Schlesinger, James, 40, 141, 147-53, 155, 157, 158, 187
single integrated operations plan (SIOP), 45, 73, 137, 150, 153-54, 156, 159
Smith, Gerard, 174
Strategic Air Command (SAC), 19, 60, 65, 71
strategic balance, 111-17
strategic doctrine: assured destruction, 137, 138, 139, 140-41, 146, 150, 153, 154, 155, 159; counterforce, 118-24, 126, 127, 137, 139, 149, 151, 155, 156-57, 158, 187, 191; damage-limitation, 136-37, 155; in the USSR, 142-47, 151-53; limited options, 46, 139, 141, 148-51, 152, 154, 159; massive retaliation, 136; sufficiency, 138-39

TERCOM, 7, 8-9, 10, 20, 45-46, 77, 158, 169-70, 171
Threshold Test Ban Treaty, 9
Triad, 43

Vance, Cyrus, 155
verification, 190-91
Vladivostok accord, 76

Zumwalt, Elmo, 32, 37

ABOUT THE AUTHOR

RON HUISKEN was awarded a Bachelor of Economics (Honours) from the University of Western Australia in 1968 and a Master of Social Sciences from the University of Stockholm in 1970. He spent a year in 1969-70 with the Stockholm International Peace Research Institute (SIPRI) followed by two years as an assistant lecturer in economics with the University of Malaya. During 1972-76 he was back at SIPRI and then spent a year as visiting fellow at the Strategic and Defense Studies Centre, Australian National University. In June 1979 he submitted his doctoral dissertation to the Department of International Relations A.N.U. and then joined the United Nations Centre for Disarmament as consultant to a group of government experts on the relationship between disarmament and development. He now works as a defense analyst with the Australian government.